Curve and Surface Fitting

An Introduction

Peter Lancaster

and

Kęstutis Šalkauskas

Department of Mathematics and Statistics
University of Calgary
Calgary, Alberta, Canada

ACADEMIC PRESS
Harcourt Brace Jovanovich, Publishers

London   San Diego   New York
Boston   Sydney   Tokyo   Toronto

ACADEMIC PRESS LTD
24–28 Oval Road
London NW1 7DX

US Edition Published by
ACADEMIC PRESS INC
San Diego, CA 92101

This book is printed on acid-free paper ©

British Library Cataloguing in Publication Data

Lancaster, P.
  Curve and surface fitting: An introduction
  1. Curve fitting
  I. Title   II. Šalkauskas, Kęstutis
  519.5′32   QA297.6

  ISBN 0-12-436060-2
       0-12-436061-0 (Pbk)

Printed in Great Britain by St Edmundsbury Press Ltd
Bury St Edmunds, Suffolk

Preface

In recent years there has been explosive growth in the computational power of computers, in the ease of access to these computers, and in computer graphics. The development of some of the techniques used in computer graphics relies on a wide range of mathematical methods for curve and surface fitting. Since access to computers requires very little training in mathematics, many of these methods may not be easily understood by the great variety of people who are now able to use powerful computing equipment. The purpose of this book is to reveal to the interested (but perhaps mathematically unsophisticated) user the foundations and major features of several basic methods for curve and surface fitting that are currently in use. In this way, the authors hope to help bridge the gap between the users and designers of curve and surface fitting methods.

The intended readership includes a great variety of users of computer graphics, such as geographers, cartographers, surveyors, geophysicists, engineers, computer scientists, and applied mathematicians. For most of the subject matter, the mathematical preparation required of the reader does not exceed the level of first-year university courses. Indeed, beginning with a review chapter of basic ideas from calculus and algebra (written in a geometrically intuitive way), the reader is brought to some understanding of progressively more advanced topics. The presentation is such that this book can be used as a reference and learning source for any of the users described above. Also, it could very well form the basis of a course for students in any of these disciplines. In fact, this book originated as lecture notes for a course given to practising geophysicists.

In some respects, the book is written in a mathematical style (reflecting the authors' profession), but we hope that the reader will not find this forbidding. We have tried to be precise and, at the same time, to be discursive and to develop a geometrical understanding of the mathematics. We have formulated important statements as theorems because we see this as a good way to give emphasis in a precise and brief form. In general, the theorems are stated without formal proof, although their meaning and the concepts involved are fully explained. From the mathematical point of view we do not strive for the greatest possible generality in the formulation of theorems (to be found elsewhere in the mathematical literature); the degree of generality is simply dictated by the level of mathematical sophistication admitted in this presentation.

The development of our subject matter has many original features and takes advantage of some unifying ideas. Almost all of the methods presented depend on the use of polynomial functions and, although the more classical methods are seen to be inflexible and inadequate in many ways, the more advanced methods using "piecewise" polynomials have extraordinary power and flexibility. Thus, there is first a progression from polynomial, to piecewise polynomial, to spline methods for curve fitting. A careful development of these univariate techniques then admits an efficient presentation of a corresponding progression for surface fitting in the later chapters.

To be more specific, Chapter 1 contains review material and summarizes the minimal information that readers need to have at their fingertips. Of course, the better prepared reader should go directly to Chapter 2 in which the classical approach to interpolation and smoothing with polynomial functions is presented. Chapter 3 is an introduction to the use of piecewise polynomial functions leading up to the notion of spline curves and their uses, which are the subject matter of Chapter 4. The presentation here is a little unusual in that natural cubic splines appear as a mix of two more primitive curve fitting methods. This "mixing" is further clarified and formalized in Chapter 5 with the introduction of projectors and Boolean sums of projectors. The approach taken to the description of cubic splines facilitates the discussion of blending and surface spline methods in subsequent chapters.

In Chapter 6 we discuss, in general terms, the criteria that might be used in comparing different methods for curve and surface fitting and then continue, in Chapter 7, with a presentation of classical techniques using polynomial methods in two variables. In Chapter 8 we show how any curve fitting technique (or pair of techniques) can be combined to generate surface fitting methods. These are the so-called tensor product and blending methods. We also show how these are conveniently described in terms of projectors.

Chapter 9 contains a presentation of finite element methods. Although finite elements were developed by engineers and numerical analysts as part of a technique for the solution of differential equations, they appear here simply as the basic units in surface construction methods. These can be seen as a generalization of some of the piecewise polynomial techniques for curve fitting introduced in Chapter 3. Least squares, "moving" least squares, and some hybrid methods are the subject matter of Chapter 10. In Chapter 11 we consider "surface splines" by which we understand bivariate methods having the same general form as spline curves, particularly with regard to their representation as Boolean sums of projectors. They are presented here in a more elementary way than can be found elsewhere in the mathematical literature.

We should conclude this description with some comment on what we do *not* attempt to do in this book. We have not been exhaustive. We have chosen to exclude more sophisticated methods such as Bernstein–Bèzier curves and surfaces, splines in tension, methods preserving monotonicity, parametric splines, isoparametric finite elements, and so on. The reader will understand these methods more readily after absorbing the relevant ideas from this work. Readers interested in further developments are referred, in the first instance, to review papers by Barnhill (1977), Boehm *et al.* (1984), Franke (1979, 1982), Lawson (1977), and Schumaker (1976).

We have also avoided reference to any but the most primitive techniques using statistics. There have recently been important developments in this area [see Wahba (1981) and Olea (1975), for example]. Although kriging is usually seen as a probabilistic method, we have been at pains to present a simplified deterministic approach to this process in Chapter 11.

We do not provide specific algorithms for obtaining fitted curves and surfaces. In view of the great variety of methods considered and the difficulty in producing reliable and portable codes, this was considered to be an unrealistic objective. However, we do give some guidance on the relative merits of different computational lines of attack. It is also the case that many software packages are readily available which incorporate the methods we describe. One of our purposes in writing the book is to equip the reader with the knowledge to make judgements on the relative merits of different methods and packages in the light of the user's own requirements.

The authors have been ably assisted by Gisele Vezina and Pat Dalgetty in the preparation of the final typescript and by M. Paolucci and J. Reddekop in computer graphics. Also, our colleague Len Bos has given invaluable assistance in programming, advice, and proofreading. Our sincere thanks go to all of them. Both authors have received some support from the Canadian Natural Sciences and Engineering Research Council for which we are duly grateful.

To Edna and Ilsè

Contents

8. Surface Interpolation by Tensor Product and Blending Methods

9. Finite Element Methods

10. Moving Least Squares and Composite Methods

11. Surface Splines

1

Functions and Graphs

1.1 Functions of a single variable

The need for fitting curves and surfaces arises principally from the fact that many physical phenomena are deemed to be continua, although our measurement of them is discrete. From such discrete information, often using the tools of mathematics, we try to reconstruct the continuum in order to learn about its features. In order to apply mathematical techniques, it is necessary to be able to describe the phenomenon quantitatively. The phenomenon in question may be the depth of a geological layer, the deflection of the wing of an aircraft under loading, the population density in an urban area, the proportion of a particular genotype in a population, and so on. We assume that the magnitude of the phenomenon, expressed as a number, depends on some underlying variables (the so-called independent variables), which can likewise be assigned numerical values. Not all of the independent variables can be taken into account; in fact, we may be unaware of some of them. For example, the elevations of the surface of the earth above some datum depend more or less continuously on the longitude and latitude, at least piecewise, and perhaps the position of the planet in its orbit and the state of its tides. For many purposes, however, it is sufficient to regard the elevation as a "function" of two independent variables representing the longitude and latitude, say. In order to make the ideas of "variable", "function", etc. more precise, we need some definitions.

We begin with some terminology and notation associated with the real-number system. The *absolute value* or *modulus* of a real number a is a measure of the distance of a from the origin measured on the real line and is denoted by $|a|$. Thus, $|2| = 2$, $|-13.5| = 13.5$, $|0| = 0$, and so on. A formal definition is given by

$$|a| = \begin{cases} a & \text{if } a \geqslant 0, \\ -a & \text{if } a < 0. \end{cases}$$

It is easily seen that for any two real numbers a and b, $|a - b|$ represents the distance between the corresponding points on the real line.

It will be necessary for us to have a precise notation for intervals of real numbers, that is, sets of real numbers corresponding geometrically to segments of the real line. The symbol $[a, b]$ with real numbers a, b satisfying $a < b$ denotes the set of all numbers, say x, for which $a \leqslant x$ and $x \leqslant b$. Such a set is a *closed* interval. In contrast, the set of all numbers x, for which $a < x$ and $x < b$, does not contain the numbers corresponding to the endpoints of the segment; it is called an *open* interval and is written (a, b).

The symbols $[a, b)$ and $(a, b]$ denote *half-open intervals* in a notation hybridized from that for closed and open intervals, and each contains just one endpoint, as the notation indicates.

We have already begun to talk about "sets" (set is an undefined primitive term), and in order to abbreviate statements about membership in a set, we introduce the symbol that means "belongs to" or "is a member of" or "is in". Thus, if S is a set of objects, then $x \in S$ reads "*x belongs to the set S*", or "*x is in S.*" For example, $x \in [1, 2]$ means that x is a real number between 1 and 2 and possibly equal to 1 or 2. Here $S = [1, 2]$. Similarly, if S is the set of all bearded persons in Inuvik, then $x \in S$ means that x is a bearded person in Inuvik.

We turn to the concept of function. Let S be a set and let a rule be given so that for every $x \in S$, a real number is uniquely assigned to x. Such a rule is called a (real valued) *function* with *domain S*. Note that the function is not a number but assigns numerical values to objects in S. It is customary to name functions by letters or groups of letters; for example, f, g, cos, exp, log. If $x \in S$, the value assigned to x by a function f is written $f(x)$ and is called the value of f at x. We will here be dealing frequently with sets S that are intervals or sets of singletons of real numbers or else with sets of pairs of real numbers. The quantities in S are values of the *independent variable*, whereas the function value is the *dependent variable*.

We consider first functions of a single independent variable in which S is an interval, which may be the whole real line. Normally, functions are defined by stating how they affect the independent variable x. We may have, for example, $f(x) = x + 1$ or $g(x) = \sqrt{x}$. In the first case, it does not matter what the domain S is, although it may be convenient to restrict x for some reason or other. In the second case, however, S may not contain any negative numbers. These are examples of algebraic functions; their values may be calculated by using a finite number of algebraic operations, such as addition, multiplication, and obtaining roots. We often refer to the *class* of algebraic functions. This class contains polynomials, rational functions (quotients of polynomials) and a multitude of others. Other familiar classes of functions are those of trigonometric functions and the so-called transcendental functions, etc. Indeed, we may classify functions according to some attributes and then use functions of a given class to accomplish a certain task. We shall introduce some classes of functions, useful for our purposes, in more detail as we progress. In curve and surface fitting, desirable attributes are continuity and smoothness. We shall therefore introduce classes of functions based on these ideas. A class of functions will generally have a common domain S, and we often need to consider combinations of such functions.

First, if f and g are functions with a common domain S, then the *sum* of the two functions is again a function, denoted by $f + g$, and its value at each $x \in S$ is defined by

$$(f + g)(x) = f(x) + g(x).$$

Next, the *product* function, written fg, is also defined on S and is given by taking the pointwise product of function values:

$$(fg)(x) = f(x)g(x),$$

for each $x \in S$. Finally, if α is any real number, we define a function written αf on S and called a *scalar multiple* of f by

$$(\alpha f)(x) = \alpha f(x),$$

for each $x \in S$. This is, of course, consistent with addition of functions in the sense that $f + f$ and $2f$ define the same function, for example.

We turn now to an introduction to the notions of continuity and smoothness.

1.2 Graphs and continuity

In order to represent a function of one real variable pictorially we make use of a graph. We set up a Cartesian coordinate system, labelling the axes x and y. The domain S of the function then corresponds to a set of points on the x-axis. The collection of all points $(x, f(x))$ as x varies over S is the *graph* of f; it may or may not be a curve. For the graph to be a smooth curve, it is, of course, necessary that S be an interval. In contrast, the curve-fitting problem begins with an S that is a set of discrete points. We form an interval, say T, containing S and look for a "reasonable" function f defined on T that comes close in some sense to the data when $x \in S$ (see Fig. 1.1). A "reasonable" function may have to be continuous and/or smooth and should not display features other than those that are known to be present in the physical event giving rise to the data.

FIG. 1.1   Domains S, T with S contained in T.

When the domain is an interval, the idea of a *continuous function* has a precise mathematical meaning. Intuitively it can be described as follows: *Let f be a function with domain S and let a ∈ S. Then f is continuous at a if the values f(x) approach the value f(a) as x approaches a from either side of a.*

In Fig. 1.2, f is continuous at a in the left-hand diagram but not in the right-hand diagram, where the value $f(a)$ is indicated by the dot and is approached by the values $f(x)$ when x approaches a from the right. However,

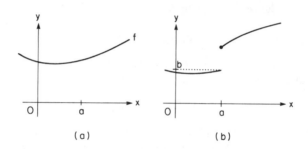

FIG. 1.2   (a) Continuity and (b) discontinuity at $x = a$.

as x approaches a from the left, $f(x)$ approaches the value $b \neq f(a)$. Incidentally, b is not a value of $f(x)$.

A more precise description of continuity lies in the following.

Definition   *Let f be a function with domain S, where S is an interval. Then f is continuous at a point a ∈ S if, given any number ε > 0, however small, there exists a number δ > 0, and depending on ε, such that whenever x ∈ S and satisfies |x − a| < δ, f(x) satisfies |f(x) − f(a)| < ε.*

Recall that the absolute value $|u - v|$ of the difference of two numbers u and v is the distance between u and v, so the definition states that the distance between $f(x)$ and $f(a)$ is less than ε whenever x is sufficiently close (within δ) to a. This definition is frequently abbreviated by introducing the term "limit" as follows: *The function f is continuous at a point x inside an interval of the domain S if* $\lim_{\xi \to x} f(\xi)$ *exists and is equal to f(x). If x is an endpoint of the interval S, the formal definition implies that the limit is* taken only from one side. Thus, if f is continuous on $[a, b]$, then the "one-sided" limits $\lim_{x \to a^+} f(\xi)$ and $\lim_{x \to b^-} f(x)$ exist and equal $f(a)$ and $f(b)$, respectively.

For practical purposes, it is sufficient to observe that discontinuities will generally arise in one of two ways:

(1) when the denominator of a quotient vanishes. For example, the function $f(x) = 1/x$ is not defined at $x = 0$. It has a discontinuity at this point.

(2) by defining f differently in parts of S, e.g. $f(x) = 0$ when $0 \leq x \leq 1$ and $f(x) = 1$ when $x > 1$. This results in f not being continuous at $x = 1$.

We shall be using (2) in subsequent chapters. An interval will be subdivided into contiguous subintervals, and a function will be constructed piecewise, subinterval by subinterval. On each subinterval, the function will be given by an expression having some "degrees of freedom", and these will be adjusted to make the total function continuous where the pieces join. If f_1 is the function in the interval S_1, and f_2 is the function in S_2, and if a is the point at which S_1 and S_2 join, we will require $f_1(a) = f_2(a)$ so as to avoid a jump there.

1.3 Derivatives and smoothness

The requirement that a function be continuous is not sufficient to make it smooth. For example, the important function f defined on all the real numbers by $f(x) = |x|$ has the graph indicated in Fig. 1.3. It is, in fact, continuous at *every* real number a, but would we wish to describe it as "smooth" at $a = 0$? The mathematical concept of smoothness involves the notion of "derivative of a function", as well as second and higher derivatives, which we shall now describe.

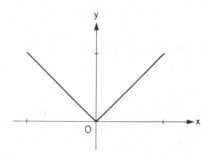

FIG. 1.3   Function $f(x) = |x|$.

Let x be a number (point) in S, the domain of f. If the graph of f does not have a sharp corner at the point with abscissa x, and f is continuous at x, we may draw a unique tangent line to the graph there and compute its slope. This may be done by calculating the quotient "rise"/"run" for the tangent, or by calculus, where we find the derivative of f and then find the value of the derivative at the number x. In any event, the slope depends on x and is therefore a function of x.

Let T be the set of points in S at which the slope of the graph of f exists and is unique:

The function whose value at $x \in T$ is the slope of the graph of f at x is called the derivative of f (or derived function of f) and is denoted by the symbol f'. The function f is said to be differentiable at each point of T.

As in the discussion of continuity, geometric intuition can be subsumed in a definition using the idea of limit as follows:

Definition   *Consider a function f whose domain is an interval S. The function f is differentiable at a point $x_0 \in S$ if $\lim_{h \to 0} h^{-1}(f(x_0 + h) - f(x_0))$ exists, where h is restricted to those values for which $x_0 + h \in S$. The derivative of f at x_0, $f'(x_0)$, is defined to be the value of this limit.*

The restriction on the permissible values of h in this definition implies that derivatives at endpoints of S, if any, are determined by appropriate one-sided limits.

Note that the set T is just the set of points in S at which the limit exists. Note also that the limit is taken of the *difference quotient*

$$\frac{f(x_0 + h) - f(x_0)}{(x_0 + h) - x_0},$$

i.e. the difference of ordinates divided by the difference of abscissas. When the derivative at x_0 exists but is not known numerically, such a quotient is frequently used as an approximation for $f'(x_0)$, provided that h is sufficiently small in absolute value. The quotient is then the slope of the chord of the graph of $y = f(x)$, as indicated by AB in Fig. 1.4 (for $h > 0$). As h decreases, we anticipate that the slope of the chord will approach the slope of the tangent at x_0.

As noted earlier, the set T is just the domain of the derived function f' of f. Other notations for the derivative of f are $f'(x)$, if we choose not to make any distinction between a function and its value at x, or $d(f(x))/dx$ or dy/dx if we set $y = f(x)$. When f is given by an explicit formula involving elementary functions, calculus often enables one to obtain a formula for $f'(x)$.

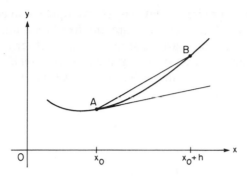

FIG. 1.4   Slope of a chord and tangent.

We may now begin to classify functions as to their continuity and smoothness over an interval $[a, b]$. Note first, however, that, since f' is a function, we may well ask whether it is continuous and whether this new function f' has a derivative at any point of its domain.

Definition $\mathcal{C}[a, b]$ *denotes the class of functions that are continuous at every point of* $[a, b]$. $\mathcal{C}^1[a, b]$ *denotes the class of functions that are continuous and have a continuous derivative at every point of* $[a, b]$.

As we shall see later, the closed interval $[a, b]$ of this definition may be replaced by an open or half-open interval, as convenient.

To illustrate, the function $f(x) = |x|$ with domain $[-1, 1]$ (see Fig. 1.3) is in the class $\mathcal{C}[-1, 1]$ but is not in the class $\mathcal{C}^1[-1, 1]$, because it has no derivative at $x = 0$; i.e. the graph does not have a unique tangent at this point. In Fig. 1.5, we sketch the graph of the function $f(x) = 1 + x^{1/2}$ on $[0, 1]$. We can say that $f \in \mathcal{C}[0, 1]$. We can also differentiate f to obtain $f'(x) = \frac{1}{2}x^{-1/2}$, and the graph of this function is also sketched in Fig. 1.5. This function is continuous at every point of $(0, 1]$, so that $f' \in \mathcal{C}(0, 1]$ and hence $f \in \mathcal{C}^1(0, 1]$. However, $f \notin \mathcal{C}^1[0, 1]$. Note that the graph of f' has a vertical asymptote at $x = 0$, and since f' is not defined there, it certainly cannot be continuous at this point.

These examples suggest that, in a certain sense, functions of class $\mathcal{C}^1$ are smoother than those that are in $\mathcal{C}[a, b]$ but not also in $\mathcal{C}^1[a, b]$. We can carry the process further to describe classes of functions that may be smoother than those of $\mathcal{C}^1[a, b]$. If the derived function f' itself has a derivative, we denote it by f'' and call it the second derivative of f. The function f'' measures the rate of change of f', the slope of the tangent, just as f' measures the rate of change of f. We denote by $\mathcal{C}^2[a, b]$ the class of continuous functions with continuous first and second derivatives at every point of $[a, b]$. These are smoother than functions that are only in $\mathcal{C}^1[a, b]$.

A primitive but very important class of examples is given by functions of the kind $f(x) = \alpha + \beta x$, where α, β are any fixed real numbers. The graph of such a function is necessarily a straight line. Furthermore, $f'(x) = \beta$, so that f' is a constant function (whose graph is a *horizontal* straight line), and $f''(x) = 0$ for all x. Thus, for any such function, we certainly have $f \in \mathscr{C}^2[a, b]$ for any interval $[a, b]$.

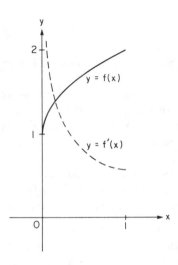

FIG. 1.5   Function $f(x) = 1 + x^{1/2}$.

An interesting example of a function that is in $\mathscr{C}^1[0, 2]$ but not in $\mathscr{C}^2[0, 2]$ is a curve made up of the following parabolic arc joined to a straight line tangent to it. This function is defined by

$$f(x) = \begin{cases} x^2, & 0 \le x \le 1, \\ 2x - 1, & 1 < x \le 2, \end{cases}$$

and its graph is sketched in Fig. 1.6. This function is certainly continuous when $x \ne 1$. As x approaches 1 from the left, the rule $f(x) = x^2$ applies, and this quantity approaches 1 as x approaches 1. When x approaches 1 from the right, we use the rule $f(x) = 2x - 1$, and again, $2x - 1$ approaches 1. Therefore the function $f \in \mathscr{C}[0, 2]$. Using calculus, we compute the function f' and obtain

$$f'(x) = \begin{cases} 2x, & 0 \le x < 1, \\ 2, & 1 < x \le 2. \end{cases}$$

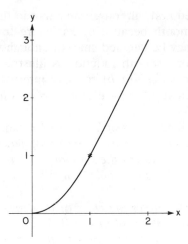

FIG. 1.6   Function of class $\mathscr{C}^1[0, 2]$. The asterisk indicates the point of discontinuity in the second derivative.

By an argument similar to that above, f' is continuous at $x = 1$, as well as at the other points of $[0, 2]$, so $f \subset \mathscr{C}^1[0, 2]$. We compute f'' now and have

$$f''(x) = \begin{cases} 2, & 0 \leqslant x \leqslant 1, \\ 0, & 1 < x \leqslant 2. \end{cases}$$

There is no way in which we can make f'' continuous at $x = 1$ because of the jump (or discontinuity) in the value of f'' at $x = 1$.

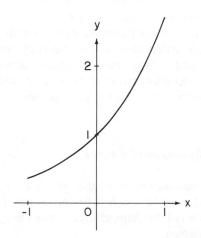

FIG. 1.7   Function $f(x) = e^x$.

It should be observed that, although the curve in this example is mathematically not very smooth because of its failure to have a continuous second derivative, it may be deemed smooth enough visually.

An example of a very smooth function is illustrated in Fig. 1.7. The function $f(x) = e^x$ is in class $\mathscr{C}[a, b]$ for any interval $[a, b]$ and has the interesting property that also $f'(x) = e^x$; i.e. $f' = f$. Consequently, $f'' = f' = f$ and so $f \in \mathscr{C}^2[a, b]$.

It is useful to define derivatives of *higher order* than 2 by the statement: *The nth derivative of f, when it exists, is the (first) derivative of the $(n - 1)$-th derivative of f and is denoted by the symbol $f^{(n)}$. Thus, $f^{(n)} = f^{(n-1)'}$.* Note that we generally write f' for $f^{(1)}$ and f'' for $f^{(2)}$.

Definition   (a) *$\mathscr{C}^n[a, b]$ denotes the class of functions that are continuous and have continuous derivatives of orders $0, 1, 2, \ldots, n$ on $[a, b]$.*

(b) *$\mathscr{C}^\infty[a, b]$ denotes the class of functions that are continuous and have continuous derivatives of all orders $0, 1, 2, \ldots$ on $[a, b]$; i.e. are infinitely differentiable.*

We often omit the $[a, b]$ in the previous notation when $[a, b]$ is obvious from the context or is all of the (extended) real line. Thus, we may write $e^x \in \mathscr{C}^\infty$. Note also that $\mathscr{C}^0$ is just $\mathscr{C}$.

It is easily verified that for each n (from 0 to ∞), the class $\mathscr{C}^n[a, b]$ has the following properties: If $f, g \in \mathscr{C}^n[a, b]$ and α is any real number, then the *sum* $f + g$ and the *scalar multiple* αf (cf. Section 1.1) are also in $\mathscr{C}^n[a, b]$. These properties of the class of functions are known as *closure* under addition and scalar multiplication, respectively, and they define the useful mathematical concept of a *vector space*. Thus, $\mathscr{C}^n[a, b]$ is the first of several examples of vector spaces that we shall encounter.

In mathematics, among the smoothest curves are those defined by functions that are infinitely differentiable. Examples of functions in $\mathscr{C}^\infty$ are e^x, $\sin x$, $\log x$ $(x > 0)$, and polynomials. Visually, however, it seems that requiring continuous second derivatives is enough, and even a continuous first derivative may suffice. We shall make extensive use of such functions in Chapters 3 and 4.

1.4 Derivatives and the shape of a curve

If the curve under consideration is the graph of a function having a continuous first and second derivative on some interval $[a, b]$ (i.e. $f \in \mathscr{C}^2[a, b]$), then some of its shape characteristics may be described using derivatives of the function.

Let us recall once more that the *first derivative* of a function f is a function f' whose value at x is the slope of the graph of f at x. The *second derivative* f'' is the function whose value at x is the slope of the graph of f' at x. In terms of *rate of change*, f' is the rate of change of f, whereas f'' is the rate of change of f', as the independent variable increases.

In order to fix these ideas geometrically, consider the following example. In Fig. 1.8 there are sketches of the graphs of f, f', f'' for the function depicted in the first sketch.

The terms increasing and decreasing must be interpreted in the context of the ordering of the real numbers, so that negative slopes are less than positive ones. In Fig. 1.8, the slope at A is positive and hence larger than the slope at B (which is negative). The function is *increasing* where its slope is *positive*, *decreasing* where the slope is *negative*, and is *stationary* where the slope is *zero*. In Fig. 1.8, one of the stationary points is at a local *maximum*, whereas the other (at C) is a local *minimum* of the function.

The function is *concave up* where the second derivative is *positive* (B to D) and *concave down* where the second derivative is negative (A to B). The changeover occurs at the *point of inflection B*; there the *second derivative has value zero*.

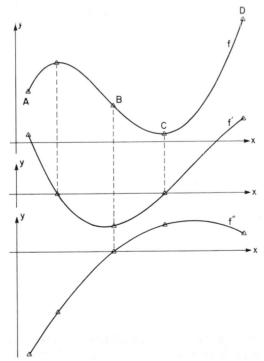

FIG. 1.8   Function with its first and second derivatives.

We conclude Section 1.4 with an introduction to the idea of "curvature", which also plays a role in discussions of the smoothness and shape of a curve. Consider any function $f \in \mathscr{C}^2[a, b]$, let x_0 be any point of $[a, b]$ and let P be the point $(x_0, f(x_0))$ (Fig. 1.9). Let C be any circle that passes through P and has a common tangent with the graph of $y = f(x)$ at P.

In some neighbourhood of P, an arc of such a circle is the graph of a function $y = g(x)$, and by construction we have $g(x_0) = f(x_0)$ and $g'(x_0) = f'(x_0)$. It can be shown that there is a *unique* circle C_0 of this type satisfying the additional condition that the second derivatives of f and g also agree at P, i.e. the condition $g''(x_0) = f''(x_0)$ holds. This circle is called the *osculating circle* at P, and its radius ρ is the *radius of curvature* at P. The reciprocal of this radius, $\kappa = 1/\rho$, is the *curvature* of the graph of f at P. Thus, the curvature is large (ρ is small) at x_0 if the tangent to $y = f(x)$ is turning "rapidly" with x as x increases through x_0. The curvature is small (ρ is large) at x_0 when the tangent is turning "slowly" as x increases through x_0. In particular, the curvature is zero (ρ is infinite) if the graph of $y = f(x)$ is a straight line.

It turns out that the curvature can be represented by

$$\kappa = |f''(x_0)|/\{1 + f'(x_0)^2\}^{3/2}. \tag{1.1}$$

The surd in the denominator is troublesome, but if it is known that $f'(x)$ is small at x_0, then we may be able to use $|f''(x_0)|$ as an approximation for the curvature.

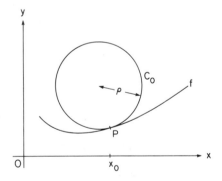

FIG. 1.9   Osculating circle.

1.5 Integration

In this section we give a heuristic introduction to the idea of *integral* and the important result known as the *Fundamental Theorem of Calculus*. A careful definition of the "definite integral" of a function requires the use of a limiting process, as in the cases of continuity and differentiability of a function at a point. The theory (due to Riemann in the nineteenth century) begins with the concept of the area of a rectangle and the idea that the definite integral of a positive function $f \in \mathscr{C}[a, b]$ represents the area between the graph of $y = f(t)$ and the t-axis, as in the hatched area of Fig. 1.10. As indicated in Fig. 1.10, the area under the curve can be *approximated* by using a set of rectangles. By carefully constructing sequences of such approximations and then applying a suitable limiting process, a math-

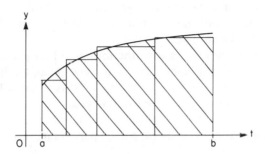

FIG. 1.10   Integral and area.

ematically viable definition is reached for a number representing the area under the graph of $y = f(t)$. This number is written as $\int_a^b f(t)\, dt$. It is, of course, to be expected intuitively that if the graph of $y = f(t)$ is reasonably smooth on $[a, b]$, then there really is a unique real number representing this area. More generally, a function f defined on (a, b) is said to be *integrable on* (a, b) if this number, $\int_a^b f(t)\, dt$, defined by the limiting process exists.

Note that where the function takes negative values in (a, b), corresponding contributions to the definite integral will also be negative. The following statements will then come as no surprise (cf. Fig. 1.3):

$$\int_{-1}^{1} t\, dt = 0, \qquad \int_{-1}^{1} |t|\, dt = 1.$$

We mention some important properties of the definite integral. First, the "range of integration" [the interval (a, b) in the previous discussion] can be broken down into subintervals. In symbols, if f is integrable on (a, b) and c is a number between a and b, then

$$\int_a^c f(t)\, dt + \int_c^b f(t)\, dt = \int_a^b f(t)\, dt. \qquad (1.2)$$

The next two properties refer to the integral of sums and scalar multiples of functions (as described in Section 1.1). If f and g are integrable on (a, b) and α is a real number, then

$$\int_a^b [f(t) + g(t)] \, dt = \int_a^b f(t) \, dt + \int_a^b g(t) \, dt, \tag{1.3}$$

$$\int_a^b \alpha f(t) \, dt = \alpha \int_a^b f(t) \, dt. \tag{1.4}$$

Taking advantage of a notion introduced in Section 1.3, the last two statements can be abbreviated to the fact that the class of all integrable functions on (a, b) forms a vector space.

It is important to note that all functions in $\mathscr{C}[a, b]$ are integrable on (a, b), but the converse statement is not true. For example, the step function defined by $f(x) = 1$ on $[0, 1]$, $f(x) = 2$ on $(1, 2)$, is integrable on $[0, 2]$ (in fact, $\int_0^2 f(t) \, dt = 3$), and f is not continuous at 1.

As we might expect, a function that is integrable on (a, b) is also integrable on any subinterval of (a, b). In particular, if $a \leqslant x \leqslant b$ and f is integrable on (a, b), then $\int_a^x f(t) \, dt$ exists. In fact, this will allow us to define a new function F on $[a, b]$ by the rule

$$F(x) = \int_a^x f(t) \, dt. \tag{1.5}$$

Now the important idea is that forming a function F from f in this way can, under widely useful hypotheses, be seen as an operation that is the inverse of the operation of forming the derived function from a function. A simple but more precise statement of this kind is the following: *if $f \in \mathscr{C}[a, b]$, then the function F defined by Eq. (1.5) is differentiable at each point of (a, b) and its derivative is f.* Thus, $F' = f$ or

$$\frac{d}{dx} \int_a^x f(t) \, dt = f(x) \qquad \text{or} \qquad \int_a^x F'(t) \, dt = F(x) - F(a).$$

In particular, if $f \in \mathscr{C}[a, b]$, then $F \in \mathscr{C}^1[a, b]$.

In other words, we can start with a function $f \in \mathscr{C}[a, b]$ and use integration to form a new function F as in Eq. (1.5); differentiation of F then returns the original function f.

A somewhat more general point of view starts with the following definition: *let $f \in \mathscr{C}[a, b]$; then any function F for which $F' = f$ on (a, b) is called an antiderivative or indefinite integral of f, and is written $F(x) = \int f(x) \, dx$.* (Note that a definite integral is a number, and an indefinite integral is a function.) Our definition [Eq. (1.5)] of a particular function F, together with the result in italics ensures that there is at least one indefinite integral.

We note that, in the standard calculus course, a lot of time is spent in studying indefinite integrals (and derived functions) of the elementary functions.

Being assured of the existence of one indefinite integral of f, it is then reassuring to find that every indefinite integral of f is closely related to the one we already have. In fact, if $f \in \mathscr{C}[a, b]$, then every indefinite integral of f differs from the function F of Eq. (1.5) by a constant function (which is generally known as "the constant of integration").

We are now in a position to remind the reader of the Fundamental Theorem of Calculus.

Theorem 1.5.1. *If $f \in \mathscr{C}[a, b]$ and F is any indefinite integral of f, then*

$$\int_a^b f(t)\, dt = F(b) - F(a).$$

This result summarizes the most important idea used in the evaluation of definite integrals. Note that because there is a *difference* of F values on the right, the same conclusion obtains whatever indefinite integral F is chosen.

1.6 Functions of two independent variables

In Sections 1.1–1.5, we have used the xy-plane for the purpose of displaying the graph of a function of a single independent variable. We may, however, use the same device for "coordinatizing" the plane, that is, for assigning to each point of the plane a pair of numbers that locate that point with respect to the x- and y-axes. The first number is the oriented distance to the point parallel to the x-axis and measured from the y-axis. The second measurement is taken parallel to the y-axis.

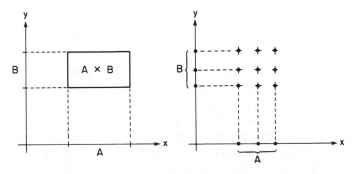

FIG. 1.11 Cartesian products.

Given a set S of points in the plane, we may have a rule assigning numerical values to the points. We then say that a function, say f, is defined on S. If P is a point in S, then $f(P)$ is the value of f at P. Previously, points were real numbers denoted by x, for example. Now P may be replaced by its number pair (x, y), and instead of $f(P)$ we write $f(x, y)$.

The set S has to be described in each case. If S consists of a number of discrete points, then its description just consists of a list. If S is a region with a boundary, then we may try to mathematically describe the boundary, and we check whether a point is in S by testing whether it lies within the boundary. There is so much flexibility in the plane, that no systematic symbolism can represent every kind of region.

However, there is a special notation for rectangular regions whose edges are parallel to the axes. Then the x-coordinate may lie between some numbers a and b; i.e. $x \in [a, b]$, whereas $y \in [c, d]$. In this case we denote S by the symbol $[a, b] \times [c, d]$, the *Cartesian product* of the two intervals. The whole plane may be thought of as an infinite rectangle, with $x \in (-\infty, \infty)$ and $y \in (-\infty, \infty)$. The notation $R = (-\infty, \infty)$ suggests the use of $R \times R$ or R^2 for the plane. This notation may be generalized to the case where x can take values in some set A of numbers, and y takes values in B. Then the points $(x, y) \in A \times B$, as in Fig. 1.11. In the second example, $S = A \times B$ contains a finite number of points, and a function on S may be defined by a list of function values. This is of course the case when the function is obtained by measurement. The problem of surface fitting then consists in taking a region containing S and finding a function on this region that agrees with the data to some extent and behaves reasonably between data points. The reasonableness often includes continuity and smoothness.

As in the case of functions of a single variable, the notion of continuity for functions of two variables can be given a precise, formal definition. Intuitively, it can be expressed by the statement:

A function f is continuous at a point P_0 in its domain S, if the values $f(P)$ approach the value $f(P_0)$ as P approaches P_0 along any path in S whatsoever.

Discontinuities often arise where division by zero can occur in the mathematical formula for the function or where a region is subdivided into patches with different formulas for the function on each patch. Such a patchwork is a very useful device and will be employed in some later chapters. Suppose that A and B are regions with common boundary C. Suppose that f is a continuous function defined on A, and g is continuous on B. We will have a continuous function on the union of A and B (see Fig. 1.12) if $f(P) = g(P)$ for every point $P \in C$.

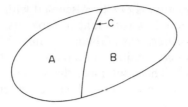

FIG. 1.12 Union of patches A and B.

The graph of a function is constructed by establishing a scale per-pendicular to the xy-plane. We usually draw a z-axis through the inter-section of the x- and y-axes, and for each $(x, y) \in S$ plot a point distance $f(x, y)$ from the coordinate plane, either above or below (x, y) depending on whether $f(x, y)$ is positive or negative. The coordinates of such a point relative to the three axes are the ordered triple $(x, y, f(x, y))$ (see Fig. 1.13).

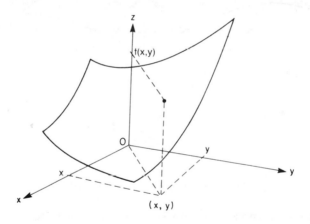

FIG. 1.13 Surface given by $z = f(x, y)$.

For the collection of all such points to be a surface, it is necessary but not sufficient for S to be a region. We say that the equation of the surface is $z = f(x, y)$.

It is important to note that strict adherence to this convention restricts the nature of the surfaces we can describe, for we admit only one value of f for each pair (x, y). To illustrate, the *whole* surface of a sphere cannot be represented in this way, but a hemisphere obtained by slicing the sphere through the equator parallel to the (x, y) coordinate plane is the graph of a function.

The question of smoothness can be discussed with reference to tangent planes. For our purposes it is sufficient to look first at slopes of tangent lines in the x- and y-directions. Given a point P in the xy-plane, we consider the two planes containing P and parallel to the xz- and yz-planes, respectively (Fig. 1.14). These intersect the surface in two curves, along which only the first (x) or second (y) of the independent variables can vary,

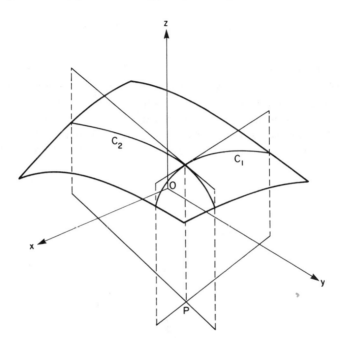

FIG. 1.14   Slope of the surface at P.

respectively. We compute the slopes of these curves as before in the case of functions of a single variable. These slopes are themselves functions of P [or (x, y), where (x, y) are the coordinates of P]. If these slope functions are continuous at some point $P_0 = (x_0, y_0)$ in the domain of f, then we can assert the following.

(1) The graph of f has a uniquely defined tangent plane at the point $(x_0, y_0, f(x_0, y_0))$,

(2) The graph of f is locally flat at the point $(x_0, y_0, f(x_0, y_0))$ in the sense that if we move away from this point in any direction whatsoever, the graph of f stays close to the tangent plane, provided only that we do not go too far. This heuristic statement can be made precise by the introduction of suitable limiting processes.

If there is no slope at a point P, the surface is not smooth there in the sense of having slope on every direction through the point P. In calculus, the slope is computed with partial derivatives, but the ratio rise/run gives the correct result if these are read off the tangent lines. Geometrically, slopes are not calculable where the graph is vertical, where there is a "fault", or where there is a sharp ridge on which the "tangent lines" can rock. In later chapters the piecewise construction of surfaces will be such that along the "seams" where patches are joined up, slopes exist and vary in a continuous manner. Since the functions will be constructed from functions of a single variable, they will inherit some smoothness from the latter.

For future purposes, we introduce the notations used in calculus to describe the slopes of the curves C_1 and C_2 defined by the point P (Fig. 1.14). Along C_1, only x varies, so the slope of C_1, where it exists, can be obtained by computing the derivative (Section 1.3) of the function f defining the surface, treating y as a constant.

Similarly, along C_2, only y varies, so we may compute the slope by computing the derivative of f, treating x as a constant. We make the following geometric definitions.

Definition *The function, whose value at point $P \in S$ is the slope of the surface defined by $z = f(x, y)$ in the x-direction evaluated at P, is called the partial derivative of f with respect to x, and is denoted by the symbol f_x. Similarly, the function whose value at P is the slope of the surface in the y-direction evaluated at P is called the partial derivative with respect to y, and is denoted by f_y.*

Other notations are f_1 and f_2 to denote derivatives with respect to the first and second variables, respectively, or $\partial f(x, y)/\partial x$ and $\partial f(x, y)/\partial y$, or $\partial z/\partial x$, $\partial z/\partial y$. These are referred to as (*partial*) *derivatives* of *order one*. More generally, we can define a *directional* derivative. This is done by considering any plane through P that is perpendicular to the xy-plane. This plane of section intersects the surface in a curve that may have a tangent at P lying in the plane of section. On the line of intersection of the xy-plane and the plane of section, we choose a positive direction and denote distance by s. Then the slope of the tangent is the directional derivative, denoted by $\partial f/\partial s$.

At points corresponding to the boundary of the domain S of the function, directional derivatives are defined by one-sided limits analogously with the situation for functions of one variable.

The first-order derivatives, when they exist, are themselves functions of two variables, and if their graphs also have slopes, these may be computed via derivatives. *Thus, if f_x has at some point P a slope both in the x- and y-directions, these are denoted variously by*

$$f_{xx} = f_{11} = \frac{\partial^2 f(x, y)}{\partial x^2} = \frac{\partial^2 z}{\partial x^2} \quad \text{and} \quad f_{xy} = f_{12} = \frac{\partial^2 f(x, y)}{\partial y \, \partial x} = \frac{\partial^2 z}{\partial y \, \partial x}.$$

Observe that the order of the subscripts is not the same as the order of the variables in the notation using ∂. Now, if f_y has slope in the x- and y-directions, then these are similarly denoted by

$$f_{yx} = f_{21} = \frac{\partial^2 f(x, y)}{\partial x \, \partial y} = \frac{\partial^2 z}{\partial x \, \partial y} \quad \text{and} \quad f_{yy} = f_{22} = \frac{\partial^2 f(x, y)}{\partial y^2} = \frac{\partial^2 z}{\partial y^2}.$$

It turns out that if f_{xy} and f_{yx} are continuous, then they are equal; then the order in which the derivatives are taken is not important. In this (very common) case, the three functions f_{xx}, f_{xy}, f_{yy} are all of the second-order (partial) derivatives of f.

For a function of two variables, this idea can be generalized to any number of derivatives with respect to x and y, as long as the function is sufficiently well behaved. Then the notation using ∂ is advantageous, provided that the order in which the operations are performed does not matter, and that is indeed the case when all the derivatives are continuous.

Then, an nth-order derivative of f, p times with respect to x and q times with respect to y, is denoted by

$$\partial^n f(x, y)/\partial x^p \, \partial y^q, \qquad p + q = n.$$

There are $n + 1$ such derivatives of order n, obtained by putting $p = 0, 1, \ldots, n$.

We are now in a position to generalize the concept of $\mathcal{C}^n[a, b]$, the class of functions (of one variable) having continuous derivatives of all orders up to and including n on the interval $[a, b]$.

In our uses of the following definitions, the region S will have a simple closed polygon as its boundary.

Definition   *Let S be a region in the xy-plane.*

(a) *$\mathcal{C}^n(S)$ denotes the class of functions that are continuous and have n continuous partial derivatives of orders $0, 1, 2, \ldots, n$ on S.*

(b) *$\mathcal{C}^\infty(S)$ denotes the class of functions that are continuous and have continuous derivatives of all orders $0, 1, 2, \ldots,$ on S, i.e. are infinitely differentiable.*

These definitions should be compared with those at the end of Section 1.3. As in the case of functions of a single variable, each class $\mathcal{C}^n(S)$ turns out to be a vector space.

1.7 Polynomial functions

In any study of curve and/or surface fitting there is one class of functions that plays a supremely important role. This is the class of polynomial functions, and we now present a brief introduction to the nature of these functions.

The main reason for their popularity is undoubtedly that it is easy to compute with them. Evaluation at a point, addition and multiplication, and differentiation and integration are all readily performed. The definition shows, in particular, that to evaluate a polynomial at a point it is only necessary to multiply and add real numbers together finitely many times.

Definition (a) *A function p defined for all real numbers x by*

$$p(x) = a_N x^N + a_{N-1} x^{N-1} + \ldots + a_1 x + a_0, \tag{1.6}$$

where N is a non-negative integer and $a_0, a_1, \ldots, a_N$ are fixed real numbers, is called a polynomial.

(b) *If p(x) has this representation and $a_N \neq 0$, then p(x) has degree N.*

(c) *If all coefficients $a_0, a_1, \ldots, a_N$ are zero, then p(x) is called the zero polynomial.*

Polynomials of degree one, two, three, . . ., are also known as linear, quadratic, cubic, . . ., polynomials, respectively. Thus $2x^2 + 1$, $(3.1) x^2 - x = (3.1) x^2 + (-1)x$ are polynomials of degree two or quadratic polynomials.

Polynomials of degree zero $(p(x) = a_0, a_0 \neq 0)$, together with the zero polynomial, are called the constant polynomials. Their graphs are straight lines parallel to the x-axis.

It will be very useful for us to be able to talk easily of certain classes, or sets, of polynomials. For this purpose, we introduce a symbol $\mathcal{P}_N$, which denotes *the set of all polynomials p with degree not exceeding N, together with the zero polynomial*. Thus, a function p is in the class $\mathcal{P}_N$, (i.e. $p \in \mathcal{P}_N$) if and only if

$$p(x) = a_N x^N + a_{N-1} x^{N-1} + \ldots + a_1 x + a_0,$$

for some real numbers $a_0, a_1, \ldots, a_N$ (and here we make no conditions on their being zero or non-zero). For example, the class $\mathcal{P}_3$ contains all constant, linear, quadratic, and cubic polynomials. Also, if $p \in \mathcal{P}_3$, then *four* parameters (coefficients in the previous expression) are needed to determine p precisely.

Polynomials can be added together in a natural way to produce a new polynomial. Thus, if $p, q \in \mathcal{P}_N$, then the sum $p + q$ (as defined in Section

1.1) is also in $\mathcal{P}_N$ and is obtained by adding corresponding terms in the expressions for p and q. Thus, if

$$p(x) = a_N x^N + a_{N-1} x^{N-1} + \ldots + a_1 x + a_0,$$

$$q(x) = b_N x^N + b_{N-1} x^{N-1} + \ldots + b_1 x + b_0,$$

then $p + q$ is defined by

$$(p + q)(x) = p(x) + q(x) = (a_N + b_N)x^N + (a_{N-1} + b_{N-1})x^{N-1}$$

$$+ \ldots + (a_1 + b_1)x + (a_0 + b_0).$$

Note that, given the degrees of p and q, say d_1 and d_2, respectively, there is, in general, little that can be said about the degree of $p + q$. The best general statement to be made is that either $p + q$ is the zero polynomial or

$$0 \leqslant \deg(p + q) \leqslant \max\{\deg p, \deg q\} = \max\{d_1, d_2\}.$$

for example (where $\Rightarrow$ means "implies"),

$$p(x) = x + 2, \qquad q(x) = -x - 2, \quad \Rightarrow \quad (p + q)(x) = 0,$$

$$p(x) = x^3 - 2x + 3, \quad q(x) = x^2 - 3, \quad \Rightarrow \quad (p + q)(x) = x^3 + x^2 - 2x.$$

For a scalar multiple (cf. Section 1.1) of a polynomial p (as above) we have that, for any real number α, the function αp is defined by

$$\alpha p(x) = \alpha(a_N x^N + \ldots + a_1 x + a_0) = (\alpha a_N)x^N + \ldots + (\alpha a_1)x + (\alpha a_0).$$

Several important properties of polynomials are summarized in the Theorem 1.7.1. The terms "vector space" and "dimension" used in the theorem have precise mathematical meanings which we will not pursue in detail. For the first term (as noted in Section 1.3) a class of functions (or vectors, or other objects) $\mathcal{S}$ with the property that $p, q \in \mathcal{S}$ implies $p + q \in \mathcal{S}$ and $\alpha p \in \mathcal{S}$ for all real numbers α, is called a *vector space*. This property is certainly enjoyed by the class $\mathcal{P}_N$ where N is any fixed positive integer. The *dimension* of $\mathcal{P}_N$ is, in effect, determined by the number of coefficients $a_0, a_1, \ldots, a_N$ needed to pin-down a particular function in $\mathcal{P}_N$.

Theorem 1.7.1   *The vector space $\mathcal{P}_N$ has dimension $N + 1$.*

Note that, in contrast to this statement, no mention is made of "dimension" in the corresponding statements concerning $\mathscr{C}^n[a, b]$ (in Section 1.3) and $\mathscr{C}^n(S)$ (in Section 1.6). The reason for this is that, in the last two examples of vector spaces, all the functions in the space cannot be described in terms of any *finite number* of parameters. Consequently, they are said to be spaces of *infinite* dimension.

Before completing this introduction, we should note that our *definition* of a polynomial function does not give a good method to evaluate a polynomial computationally. If we work directly with the definition, the powers of x up to the Nth would be computed, multiplied by the coefficients a_j, and then summed. This is found to take $\frac{1}{2}N(N+1)$ multiplications and N additions. In contrast, *Horner's scheme* requires only N multiplications and N additions. This scheme can be described by introducing multiple parentheses in the expression for p that indicate the order in which operations are to be performed:

$$p(x) = (\ldots((a_N x + a_{N-1})x + a_{N-2})x + \ldots + a_1)x + a_0. \qquad (1.7)$$

1.8. Polynomials as products

Polynomials can also be multiplied together in the familiar term-by-term fashion. Thus,

$$p(x) = x - 1, \qquad q(x) = x + 3, \quad \Rightarrow \quad (pq)(x) = p(x)q(x)$$
$$= x^2 + 2x - 3,$$

$$p(x) = (0.1)x^4 - (1.8)x, \quad q(x) = 2, \qquad \Rightarrow \quad (pq)(x) = p(x)q(x)$$
$$= (0.2)x^4 - (3.6)x.$$

If f is a function with domain S and the number $x_0 \in S$ is such that $f(x_0) = 0$, then x_0 is called a *zero* of f, or a *root of the equation* $f(x) = 0$. Clearly, if f is continuous on S, then the graph of f crosses or touches the x-axis at x_0. For a polynomial p of degree one, we can write $p(x) = ax + b$ for some numbers a, b with $a \neq 0$. Such a polynomial [defined on $(-\infty, \infty)$] has one and only one zero, namely at $x_0 = -b/a$.

Consider two such polynomials of the form

$$p_1(x) = x - x_1, \qquad p_2(x) = x - x_2.$$

The product polynomial, say q, is in $\mathcal{P}_2$ and has the form

$$q(x) = (x - x_1)(x - x_2) = x^2 - (x_1 + x_2)x + x_1 x_2.$$

Thus, q has zeros at x_1 and x_2, where x_1, x_2 are the zeros of p_1, p_2, respectively. This situation is shown in Fig. 1.15. More generally, if $x_1, \ldots, x_N$ are real numbers, then we can define a polynomial $q \in \mathcal{P}_N$ by

$$q(x) = (x - x_1)(x - x_2) \ldots (x - x_N) = \prod_{i=1}^{N} (x - x_i),$$

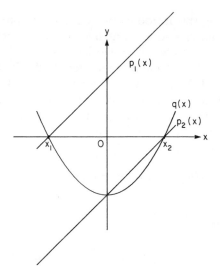

FIG. 1.15 Product of two first-degree polynomials.

and q will inherit the zeros $x_1, x_2, \ldots, x_N$ from its linear factors. This point is illustrated in Fig. 1.16, where the graphs of three polynomials from $\mathscr{P}_3$ are shown.

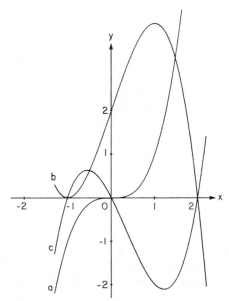

FIG. 1.16 Three cubic polynomials on $[-1.3, 2.2]$. (a) x^3, (b) $-x^3 + 3x + 2$, and (c) $x^3 - x^2 - 2x$.

We have been considering the construction of polynomials of degree N as products of N polynomials of degree one (i.e. of linear polynomials). Can every polynomial of degree N with real coefficients be expressed in this way? It is very important to realize that the answer is *No*—not if we confine our attention to factors that also have *real* coefficients.

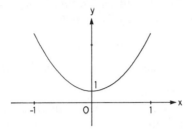

FIG. 1.17   Quadratic polynomial with no real zeros.

To illustrate this, consider the quadratic $p(x) = x^2 + 1$. Figure 1.17 shows that $p(x)$ has no real zero, and this suggests immediately that $p(x)$ cannot be factored into the product of two products of two linear polynomials; otherwise the graph would have to meet the x-axis somewhere. The same conclusion is reached when one tries to solve the equation $x^2 + 1 = 0$ by the formula method.

In order to express every real polynomial of degree N as a product of N polynomials of degree one, it is necessary to extend the concept of number to admit the complex numbers. Then it can be done. For example, $x^2 + 1 = (x - i)(x + i)$. This is a consequence of the *Fundamental Theorem of Algebra*, which states that every polynomial of positive degree with real or complex coefficients has at least one (generally complex) zero. Once this important fact is given, it is relatively easy to show that every (complex) polynomial of degree $N > 0$ has N (generally complex and possibly multiple) zeros. In Section 1.9, we investigate a little more closely the relationship between the zeros of a polynomial and the existence of linear factors.

1.9 The division process and zeros of polynomials

The division of one polynomial by another is a fundamental idea of algebra, and a technique for performing such a process is included in the high school curriculum. Given polynomials $a(x)$ and (non-zero) $b(x)$, the objective is to find a *quotient* polynomial $q(x)$ and a *remainder* polynomial $r(x)$ whose degree is less than that of $b(x)$, or is the zero polynomial, for which

$$a(x)/b(x) = q(x) + [r(x)/b(x)],$$

or, which is the same, for which $a(x) = q(x)b(x) + r(x)$. A process for finding $q(x)$ and $r(x)$ is called a *division algorithm*.

In order to clarify the significance of this process, the following result is needed and can be found in college or first-year university texts on algebra.

Theorem 1.9.1 *Let $a(x)$, $b(x)$ be polynomials, $b(x) \neq 0$. Then there exist unique polynomials $q(x)$ and $r(x)$ for which*

$$a(x) = q(x)b(x) + r(x),$$

and either $r(x) = 0$ or $r(x)$ has degree less than that of $b(x)$.

A particularly important case of this result is that in which $b(x) = x - x_0$ for some fixed real number x_0 – a polynomial of degree one. Then the result says that there are unique polynomials $q(x)$ and $r(x)$ for which

$$a(x) = q(x)(x - x_0) + r(x),$$

where $r(x) = 0$ or $r(x)$ has degree zero; i.e. in either case, $r(x)$ is a constant polynomial, say r. Thus,

$$a(x) = q(x)(x - x_0) + r.$$

If we now substitute x_0 for x, it is found that $r = a(x_0)$. This is the gist of another important result known as the *remainder theorem*.

Theorem 1.9.2 *The remainder on division of the polynomial $a(x)$ by $x - x_0$ is $a(x_0)$.*

The following *factor theorem* is an immediate consequence of the remainder theorem:

Corollary 1.9.3 *The number x_0 is a zero of the polynomial $a(x)$ if and only if $a(x)$ has a factor $x - x_0$.*

Thus, x_0 is a zero of the polynomial $a(x)$ if and only if $a(x) = q(x)(x - x_0)$ for some polynomial $q(x)$. Recall that there may be *no* real x_0 that is a zero of a given polynomial. This is the case for polynomials $2x^2 + 3$ and 1 (the constant polynomial), for example. In order to make a more general statement about the existence of a zero of a polynomial, it is necessary (as in Section 1.8) to introduce the study of complex numbers and polynomials whose coefficients are complex numbers. We shall not go into this.

Suppose that $a(x)$ has a zero at x_0. Then, $a(x) = q(x)(x - x_0)$. It is possible that the quotient polynomial $q(x)$ also has a zero at x_0, and it is important for us to introduce some terminology for this situation. Thus, we say that x_0 is a *simple zero* of $a(x)$ if $a(x) = q(x)(x - x_0)$ and $q(x_0) \neq 0$; i.e. x_0 is *not* a zero of the quotient $q(x)$.

On the other hand, if x_0 is a simple zero of $q(x)$ [as well as being a zero of $a(x)$], then we have $q(x) = q_1(x)(x - x_0)$ with $q_1(x_0) \neq 0$ and $a(x) = q(x)(x - x_0)$. Substituting for $q(x)$ from the first equation, we deduce that

$$a(x) = q_1(x)(x - x_0)^2,$$

and $q_1(x_0) \neq 0$. In this case, we say that $a(x)$ has a *double zero* at x_0.

More generally, if $a(x) = q(x)(x - x_0)^m$, where $q(x)$ is a polynomial with $q(x_0) \neq 0$, we say that $a(x)$ has a *zero of order m* at x_0. If $m > 1$, then $a(x)$ is said to have a *multiple zero* at x_0.

1.10 Graphs of polynomials

We shall see many examples of graphs of polynomials in the sequel. In Section 1.10, a few points are made that are generally useful in sketching or reading such graphs.

(1) The graph of every polynomial of degree one is a straight line. Polynomials of degree zero, i.e. constant polynomials, also have straight-line graphs; these are parallel to the x-axis (horizontal).

(2) The graph of every polynomial of degree two is a parabola with axis of symmetry vertical. If $a_2 > 0$, the parabola opens upward; if $a_2 < 0$, it opens downward (cf. Fig. 1.18).

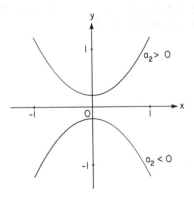

FIG. 1.18 Parabolas.

(3) It is a consequence of the Fundamental Theorem of Algebra that the graph of a polynomial of degree N can cross the x-axis at most N times; i.e. a polynomial of degree N has at most N zeros. Consequently, graphs of high-degree polynomials can display numerous oscillations.

(4) A polynomial of degree N behaves very much like its leading term $a_N x^N$ when x is large and positive or negative. As a consequence, polynomials become large in value when x becomes large (+ or −). This phenomenon is illustrated in Fig, 1.19.

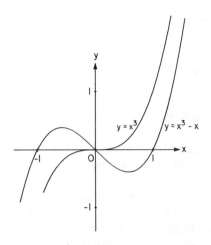

FIG. 1.19   Cubic polynomial and its leading term.

(5) A polynomial $p(x)$ *crosses* the x-axis at $x = c$ only if $p(x)$ has a zero of odd order at c (Section 1.9). If $p(x)$ has a factor $(x - c)$, then the graph of $p(x)$ has a *horizontal tangent* at $x = c$ if and only if c is a multiple zero of $p(x)$ (see Section 1.9 and Fig. 1.19).

2

Curve Fitting with Polynomials

2.1 Polynomial interpolation

The most primitive and important class of functions for the purpose of curve fitting is undoubtedly the class of polynomial functions introduced in Sections 1.7–1.10. The most elementary notion of curve fitting is that of interpolation, and so we begin by combining these with a discussion of polynomial interpolation.

In applications, we frequently come across a set of data that is thought to be derivable from some underlying function that also fills in the gaps between the data in some way. More precisely, we often have a set of function values (e.g. elevations) at some points, perhaps slopes at the same or other points, and wish to find the function that fits this data. This problem is still not sufficiently well stated. In the first place, the meaning of "fits" is not given, and if this were to be clarified, then "the function" is most likely far from unique. Let us examine a situation where this happens. Let two distinct x-values x_0 and x_1 be given together with corresponding function values f_0 and f_1. There exists a unique first-degree polynomial $a_1 x + a_0$ that fits the data in the sense that $a_1 x_0 + a_0 = f_0$ and $a_1 x_1 + a_0 = f_1$; i.e. its graph is a straight line passing through the points (x_0, f_0) and (x_1, f_1). However, there are infinitely many second-degree polynomials (parabolas) that fit the same data in the same way. In order to clarify this issue, we must state at least

(1) what concept of fit we are using, and
(2) what kinds of functions we shall use to achieve this fit.

If (1) and (2) are properly chosen, there may exist a unique function of the kind allowed in (2), achieving the type of fit required in (1).

In the case of the simplest kind of *polynomial interpolation*, the data consist of distinct x-values and corresponding f-values; let us denote them by

$$\{x_0, x_1, \ldots, x_N\}, \quad \{f_0, f_1, \ldots, f_N\}.$$

As illustrated above, the fit consists of the requirement that the function $f(x)$ employed *interpolate*; that is, it must satisfy

$$f(x_i) = f_i, \quad i = 0, 1, \ldots, N.$$

This means that the graph of $f(x)$ must pass through the points (x_i, f_i), $i = 0, 1, \ldots, N$, in the xy-plane. The functions permitted are those in $\mathcal{P}_N$, i.e. polynomials of degree N or less. In this situation there is a fundamental result that shows that we have made a happy choice for (1) and (2).

Theorem 2.1.1   *If $\{x_0, x_1, \ldots, x_N\}$ is a set of $N + 1$ distinct numbers then, for any set of numbers $\{f_0, f_1, \ldots, f_N\}$ there exists a unique polynomial in the class $\mathcal{P}_N$, say $p(x)$, such that $p(x_i) = f_i$, $i = 0, 1, \ldots, N$; i.e. whose graph passes through the $N + 1$ distinct points (x_i, f_i), $i = 0, 1, \ldots, N$.*

Note carefully that the polynomial has degree N or less. Thus, if the $N + 1$ data points lie on a straight line, the polynomial in $\mathcal{P}_N$ will look like

$$0x^N + 0x^{N-1} + \ldots + 0x^2 + a_1 x + a_0,$$

where a_1 and a_0 have appropriate values.

Note also that the general Nth-degree polynomial contains $N + 1$ constants $a_0, \ldots, a_N$ that have to be found. The theorem says that the $N + 1$ interpolation conditions determine these uniquely.

As an example, consider first the geometrically obvious case of just two points. Let $x_0 = 0$, $x_1 = 1$; $f_0 = 1$, $f_1 = 3$. Since there are two distinct points ($N + 1 = 2$), $N = 1$, and so there exists a unique first-degree polynomial that interpolates. Let it be $p(x) = a_1 x + a_0$. Imposition of the interpolation conditions yields the two equations $a_0 = 1$ and $a_1 + a_0 = 3$. Solving, we get $a_0 = 1$, $a_1 = 2$, and so $p(x) = 2x + 1$.

Next, consider polynomial interpolants as approximations for the function $1/(1 + x^2)$ on the interval $[-4, 4]$. For $N = 3, 6, 12, \ldots$, we choose for the $N + 1$ abscissas x_i equally spaced points placed on the interval $[-4, 4]$, with $x_0 = -4$ and $x_N = 4$. Then, for each x_i take $f_i = 1/(1 + x_i^2)$. The resulting polynomial interpolants of degrees 3, 6, and 12 are shown in Fig. 2.1.

It will be observed that, as N increases, the approximation seems to improve near the centre of the interval, although oscillatory behaviour is pronounced, particularly at the ends of the interval.

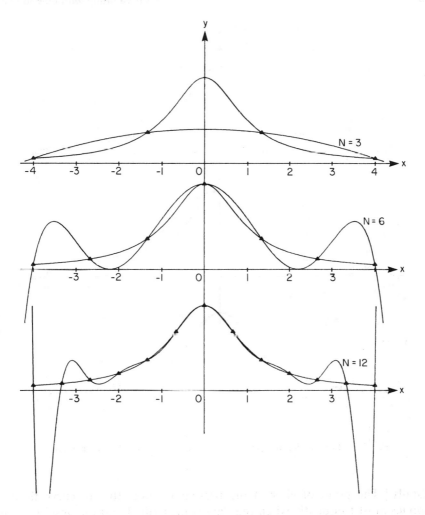

FIG. 2.1 Polynomial interpolants.

   This example suggests that, if the x_i are free for choice, then one may
do better in the approximation process if, instead of choosing evenly spaced
abscissas, we allow them to cluster near the ends of the interval. This is
indeed the case. In Fig. 2.2 we show comparable results where, for each
N, the numbers $x_0, x_1, \ldots, x_N$ are proportional to the zeros $z_0, z_1, \ldots, z_N$
(known to be all real and distinct) of the Chebyshev polynomial of degree
$N + 1$. For convenience, the constant of proportionality is chosen such that
$x_0 = -4$ and $x_N = 4$. [We refer the interested reader to any standard text
on Numerical Analysis, such as Conte and de Boor (1980) for further

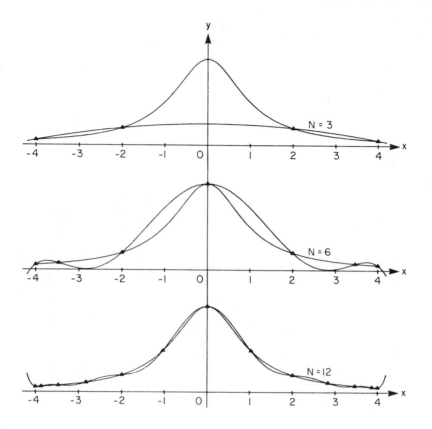

FIG. 2.2   Polynomial interpolants at the (extended) Chebyshev abscissas.

details.] The point of view of the present volume, that of curve fitting, applies most frequently when the data is prescribed and the abscissas are *not* free for choice. The reduction in error of approximation by the choice of abscissas is a topic of approximation theory that is not investigated here.

We next introduce some examples that will be used repeatedly throughout this volume in order to facilitate comparisons between different methods of curve (and, subsequently, surface) fitting. The data for curve-fitting procedures are sampled from mathematically defined functions that can be thought of, in a picturesque way, as a "mountain" on a plain together with a "ramp/mountain" combination. The graphs of these functions are common to the several parts of Figs. 2.3 and 2.4, respectively.

In Fig. 2.3, we illustrate polynomial interpolants of degrees 4, 8, and 16 for the mountain example when abscissas are equally spaced (on the left)

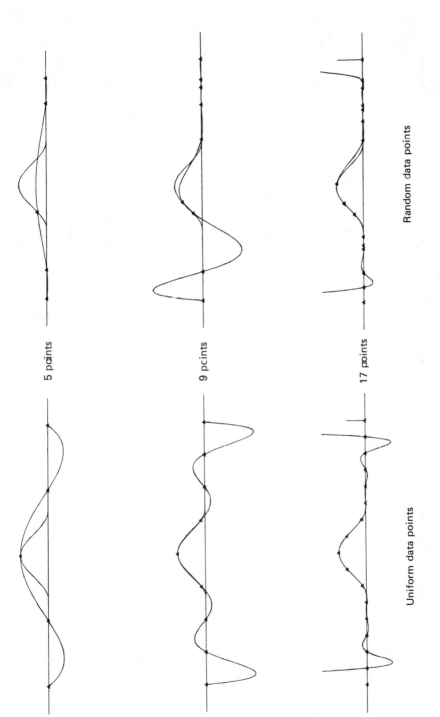

5 points

9 points

17 points

Uniform data points

Random data points

FIG. 2.3   Polynomial interpolants: mountain.

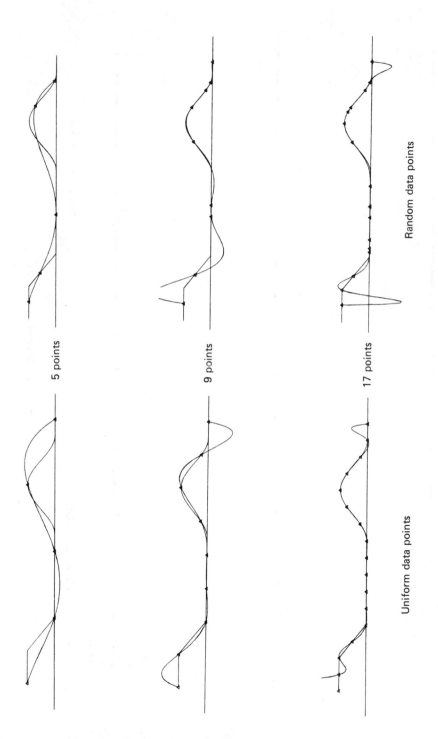

Uniform data points   Random data points

5 points

9 points

17 points

Fɪɢ. 2.4   Polynomial interpolants: ramp plus mountain.

and randomly selected (on the right). The irregular spacing is likely to be representative of many practical situations. In Fig. 2.4, comparable results are illustrated for the ramp/mountain case. The sets of abscissas used are just those of Fig. 2.3. The oscillatory behaviour and large swings in value of the higher degree polynomials are discouraging and indicate that "high degree" is not to be equated with "good". Consider the comments in (3) and (4) of Section 1.10.

2.2 Remarks on the computation of polynomial interpolants

Let us consider briefly the numerical implementation of the theorem stated in Section 2.1. The most natural way of finding the value of the interpolating polynomial p at an arbitrary value of x is first to find the coefficients a_0, a_1, ..., a_N in the definition of p (Section 1.7) and then to evaluate $p(x)$ by the Horner scheme [cf. Eq. (1.7)].

To find the coefficients of p we have the $N+1$ conditions $p(x_i) = f_i$, $i = 0, 1, \ldots, N$, and if we write these out in full, we have

$$
\begin{aligned}
a_0 + a_1x_0 + a_2x_0^2 + \ldots + a_Nx_0^N &= f_0, \\
a_0 + a_1x_1 + a_2x_1^2 + \ldots + a_Nx_1^N &= f_1, \\
&\vdots \\
a_0 + a_1x_N + a_2x_N^2 + \ldots + a_Nx_N^N &= f_N.
\end{aligned}
\tag{2.1}
$$

It is customary, useful, and convenient to write this system of $N+1$ equations in the $N+1$ unknowns, $a_0, a_1, \ldots, a_N$ in matrix form. We define the following matrices:

$$
V = \begin{bmatrix}
1 & x_0 & x_0^2 & \cdots & x_0^N \\
1 & x_1 & x_1^2 & \cdots & x_1^N \\
& & \vdots & & \\
1 & x_N & x_N^2 & \cdots & x_N^N
\end{bmatrix}, \quad
\mathbf{a} = \begin{bmatrix} a_0 \\ a_1 \\ \vdots \\ a_N \end{bmatrix}, \quad
\mathbf{f} = \begin{bmatrix} f_0 \\ f_1 \\ \vdots \\ f_N \end{bmatrix}, \tag{2.2}
$$

(the last two are also known as column vectors), and the set of equations can then be abbreviated to $V\mathbf{a} = \mathbf{f}$, where V, $\mathbf{f}$ are known, and $\mathbf{a}$ is to be found.

Now the computational task of solving equations in this form is a familiar one. The fact that the abscissas are distinct means that the inverse matrix V^{-1} exists and (as our theorem asserts) there is a unique solution to the equations, which can be written $\mathbf{a} = V^{-1}\mathbf{f}$.

Every computer installation that has a commitment to scientific work will have a package program available for this task, and as long as N is not too large and the x_i's not too close together, the package program will doubtless give a sufficiently accurate solution. However, the coefficient matrix V, known as a Vandermonde matrix, is notorious. For modest sizes of N, it is very ill-conditioned (in particular, its determinant is very small), and solutions may be subject to severe accumulation of round-off error, or the program may fail altogether. Thus, this direct algebraic attack on the problem is *not* recommended for a general-purpose computer code.

Note also that, for essentially the same reason, when some abscissas are very close together, then the polynomial interpolant, even when accurately computed, will be very sensitive to errors in the corresponding values f_i. This is a geometrically obvious phenomenon.

Thus, although theoretically important, this natural approach to evaluating p is not suitable for large-scale computation. The process to be described next in Section 2.3 not only has considerable theoretical interest, it is also computationally much more satisfactory. From the strictly computational point of view, however, even this is not the best. The interested reader is referred to appropriate algorithms, depending on divided difference representations for some strong competitors [and to be found in texts on numerical analysis, such as Conte and de Boor (1980)].

2.3 Lagrange's method for interpolation

In this section we describe a method attributed to the French mathematician Lagrange (1736–1813) for finding the polynomial interpolant $p \in \mathscr{P}_N$ to the data pairs (x_i, f_i), $i = 0, 1, \ldots, N$. Instead of calculating the coefficients a_0, $a_1, \ldots, a_N$ of p, we begin by considering a set of $N + 1$ different problems, each of which is relatively easy to solve and which can subsequently be combined to solve the problem posed initially.

The idea is to solve the $N + 1$ simple interpolation problems obtained by taking the distinct abscissas $x_0, x_1, \ldots, x_N$ as already presented, but taking the $N + 1$ primitive sets of f values,

$$\{1, 0, \ldots, 0\} \quad \{0, 1, 0, \ldots, 0\}, \quad \ldots, \quad \{0, \ldots, 0, 1\} \qquad (2.3)$$

in turn. According to the basic theorem (Theorem 2.1.1), each of these problems has a unique solution.

Before considering the general case, let us examine a primitive example with just two data points $(x_0, f_0) = (0, 1)$ and $(x_1, f_1) = (1, 3)$.

There are now two primitive sets of f-values:

$$\{1, 0\} \quad \text{and} \quad \{0, 1\}.$$

We solve first for the polynomial in $\mathcal{P}_1$ passing through $(0, 1)$ and $(1, 0)$. We denote this by L_0. It is found that $L_0(x) = -x + 1$ [check that $L_0(x_0) = 1$ and $L_0(x_1) = 0$; subscripts identical corresponds to the value 1, subscripts different corresponds to the value 0].

Then, we solve for the polynomial L_1 in $\mathcal{P}_1$ passing through $(0, 0)$ and $(1, 1)$. It is found that this polynomial is $L_1(x) = x$ [check again that $L_1(0) = 0$ and $L_1(1) = 1$].

It is now clear that the original interpolation problem is solved by the simple linear combination of L_0 and L_1 given by

$$p(x) = f_0 L_0(x) + f_1 L_1(x) = 1(-x + 1) + 3(x) = 2x + 1.$$

In fact, we can say more. For any pair of ordinates f_0, f_1, the polynomial in $\mathcal{P}_1$ given by

$$p(x) = f_0 L_0(x) + f_1 L_1(x).$$

solves the interpolation problem with data $(0, f_0)$ and $(1, f_1)$ because

$$p(x_0) = f_0 L_0(x_0) + f_1 L_1(x_0) = f_0(1) + f_1(0) = f_0,$$
$$p(x_1) = f_0 L_0(x_1) + f_1 L_1(x_1) = f_0(0) + f_1(1) = f_1.$$

The functions L_0, L_1 are known as the *fundamental Lagrange polynomials* determined by the pair of abscissas 0 and 1. They are also called the *cardinal functions* for polynomial interpolation at these two abscissas.

Let us return now to the general case in which $N + 1$ distinct abscissas $x_0, x_1, \ldots, x_N$ are given. Using the $N + 1$ data sets [Eq. 2.3], we are to construct $N + 1$ cardinal functions $L_0, L_1, \ldots, L_N$ in the class $\mathcal{P}_N$ with the properties

$$L_i(x_j) = \begin{cases} 1, & i = j, \\ 0, & i \neq j. \end{cases} \tag{2.4}$$

When these are found, the original problem is solved by defining

$$p(x) = f_0 L_0(x) + f_1 L_1(x) + \ldots + f_N L_N(x) = \sum_{i=0}^{N} f_i L_i(x), \tag{2.5}$$

for, using Eq. (2.4), we have $p(x_i) = f_i$, $i = 0, 1, \ldots, N$, and since $L_0, L_1, \ldots, L_N$ are in $\mathcal{P}_N$ and $\mathcal{P}_N$ is a vector space, it follows that p is also in $\mathcal{P}_N$.

It only remains to find the cardinal functions, and this can be done explicitly. Note that, by Eq. (2.4), L_i has N distinct zeros at $x_0, x_1, \ldots, x_{i-1}, x_{i+1}, \ldots, x_N$. As pointed out in Section 1.9, this means that $L_i(x)$ has factors $(x - x_0)$, $(x - x_1)$, and so on. Since L_i has degree at most N, it follows that we may write

$$L_i(x) = k(x - x_0)(x - x_1) \ldots (x - x_{i-1})(x - x_{i+1}) \ldots (x - x_N)$$

$$= k \prod_{\substack{j=0 \\ j \neq i}}^{N} (x - x_j), \tag{2.6}$$

for some real number k. To evaluate k, we use the remaining piece of data about L_i, namely that $L_i(x_i) = 1$. Thus,

$$L_i(x_i) = 1 = k \sum_{\substack{j=0 \\ j \neq i}}^{N} (x_i - x_j).$$

Solve for k and substitute back into Eq. (2.6) to obtain the result

$$L_i(x) = \frac{(x - x_0) \ldots (x - x_{i-1})(x - x_{i+1}) \ldots (x - x_N)}{(x_i - x_0) \ldots (x_i - x_{i-1})(x_i - x_{i+1}) \ldots (x_i - x_N)}$$

$$= \prod_{\substack{j=0 \\ j \neq i}}^{N} (x - x_j) \bigg/ \prod_{\substack{j=0 \\ j \neq i}}^{N} (x_i - x_j), \qquad i = 0, 1, \ldots, N. \tag{2.7}$$

Note that the factor $x - x_i$ is missing from the numerator and $x_i - x_i$ from the denominator.

EXAMPLE   The cardinal functions for polynomial interpolation at the three abscissas $x_0 = 0$, $x_1 = 1$, $x_2 = 3$ are [using Eq. (2.7)] as follows:

$$L_0(x) = \frac{(x - 1)(x - 3)}{(0 - 1)(0 - 3)} = \tfrac{1}{3}(x^2 - 4x + 3),$$

$$L_1(x) = \frac{(x - 0)(x - 3)}{(1 - 0)(1 - 3)} = -\tfrac{1}{2}(x^2 - 3x),$$

$$L_2(x) = \frac{(x - 0)(x - 1)}{(3 - 0)(3 - 1)} = \tfrac{1}{6}(x^2 - x).$$

The interpolating polynomial for function values f_0, f_1, f_2 at 0, 1, 3, respectively, is then given by Eq. (2.5):

$$p(x) = f_0 \tfrac{1}{3}(x^2 - 4x + 3) - f_1 \tfrac{1}{2}(x^2 - 3x) + f_2 \tfrac{1}{6}(x^2 - x).$$

The three cardinal functions are shown in Fig. 2.5.

It is clear from Eq. (2.7) that each cardinal function has degree N ($N = 2$ in the example). However, if the data points should lie on a straight line, then the summation in Eq. (2.5) will have cancellations in such a way that the interpolating polynomial p will be in $\mathcal{P}_1$.

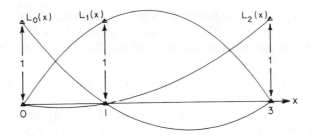

FIG. 2.5   Three cardinal functions: $N = 2$.

In fact, it can be proved that *every* polynomial in $\mathcal{P}_N$ can be expressed in the form of Eq. (2.5). The summation in Eq. (2.5) is described as a *linear combination* of the functions $L_0, L_1, \ldots, L_N$, so in other words, every polynomial in $\mathcal{P}_N$ is some linear combination of the fundamental Lagrange

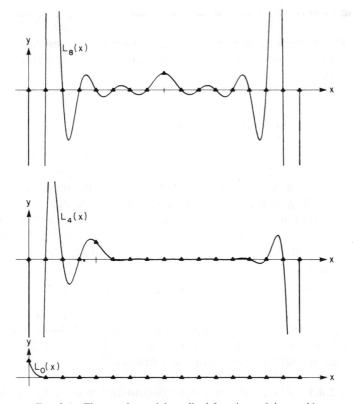

FIG. 2.6   Three polynomial cardinal functions of degree 16.

polynomials. Since these functions enjoy another technical property known as linear independence, they are said to form a *basis* for $\mathcal{P}_N$ (a *cardinal basis* in this case). This particular basis has the defining property that if

$$p(x) = \sum_{i=0}^{N} b_i L_i(x), \text{ then } p(x_j) = b_j, j = 0, 1, \ldots, N.$$

If we return to the defining expression [Eq. (1.6)] for a member p of $\mathcal{P}_N$, we observe that p is expressed there as a linear combination of $N + 1$ other functions. In this case, the functions in question are the *power functions* given by $1, x, x^2, \ldots, x^N$. These are also linearly independent, and so they also form a basis for $\mathcal{P}_N$. (Indeed, any non-trivial real vector space will always have infinitely many bases.) It is important to realize that, although Eqs. (1.6) and (2.5) may look very different, any given function in $\mathcal{P}_N$ can be represented in either way.

In Fig. 2.6, we illustrate some polynomial cardinal functions associated with 17 equally spaced abscissas. They show how the presence of data at the point where the function has the value 1 is incorporated into the complete polynomial $p(x)$. Here, the $L_i(x)$ terms oscillate and grow as $|x|$ grows. This behaviour demonstrates again (what are often seen as) some unfortunate characteristics of high-degree polynomial interpolation (cf. the examples of Section 2.1).

2.4 Hermite or osculatory interpolation

The method by which the polynomial interpolation problem was solved in Section 2.1 is attributed to Lagrange. The name of Hermite (1822–1901) is associated with a more general polynomial interpolation problem that can also be solved by means of cardinal functions, although there are other methods. In its simplest form, the data is given at $N + 1$ distinct points x_0, $x_1, \ldots, x_N$ as before, and at these we are given not only ordinates f_0, f_1, $\ldots, f_N$ but also slopes $f_0', f_1', \ldots, f_N'$. We are to find a polynomial $p(x)$ with the properties

$$p(x_i) = f_i, \qquad p'(x_i) = f_i', \qquad i = 0, 1, \ldots, N.$$

There are $2(N + 1) = 2N + 2$ conditions here, so $p(x)$ will require that many degrees of freedom [i.e. coefficients in Eq. (1.6)]—its degree must be $2N + 1$. It turns out that the function class $\mathcal{P}_{2N+1}$ is, in fact, a happy match with this set of interpolating constraints.

Theorem 2.4.1 *There exists a unique polynomial in class $\mathcal{P}_{2N+1}$ satisfying the above conditions.*

The full generality of this theorem is seldom used in computational practice. So, for the purpose of illustration, and for future reference, we restrict attention to the case $N = 1$, i.e. two points only. There are four pieces of data, two at each point, and the degree of $p(x)$ may be as high as three. To solve this problem, we first construct four cardinal functions H_0, K_0, H_1, K_1. These are determined by Hermite interpolation at the points x_0 and x_1 with four primitive data sets. In fact, we impose the interpolatory conditions

$$H_0(x_0) = 1, \qquad K_0(x_0) = 0, \qquad H_1(x_0) = 0, \qquad K_1(x_0) = 0,$$
$$H_0'(x_0) = 0, \qquad K_0'(x_0) = 1, \qquad H_1'(x_0) = 0, \qquad K_1'(x_0) = 0,$$
$$H_0(x_1) = 0, \qquad K_0(x_1) = 0, \qquad H_1(x_1) = 1, \qquad K_1(x_1) = 0,$$
$$H_0'(x_1) = 0, \qquad K_0'(x_1) = 0, \qquad H_1'(x_1) = 0, \qquad K_1'(x_1) = 1.$$

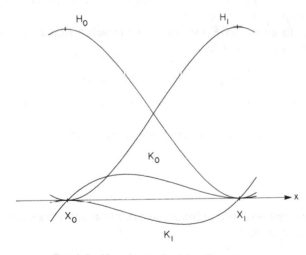

FIG. 2.7   Hermite cardinal functions; $N = 1$.

Each of these cardinal functions is a cubic and is not hard to construct. The graphs are shown in Fig. 2.7. The interpolating polynomial is then

$$p(x) = f_0 H_0(x) + f_0' K_0(x) + f_1 H_1(x) + f_1' K_1(x). \tag{2.8}$$

In order to construct the cardinal functions, observe that each is a polynomial of degree three. Now H_0 has a double zero at $x = x_1$ (cf. Section 1.9), so

$$H_0(x) = a(x + b)(x - x_1)^2,$$

where a and b still have to be determined from the conditions $H_0(x_0) = 1$ and $H_0'(x_0) = 0$. These yield

$$a(x_0 + b)(x_0 - x_1) = 1,$$
$$a(x_0 - x_1)^2 + 2a(x_0 + b)(x_0 - x_1) = 0.$$

From the second equation, we may cancel out a and $(x_0 - x_1)$ to get

$$(x_0 - x_1) + 2(x_0 + b) = 0,$$

and so $b = (x_1 - 3x_0)/2$. Substitute this into the first equation and solve for a:

$$a = \frac{1}{(x_0 - x_1)^2[x_0 + \frac{1}{2}(x_1 - 3x_0)]} = \frac{2}{(x_0 - x_1)^2(x_1 - x_0)} = -\frac{2}{(x_0 - x_1)^3}.$$

If we put $x_1 - x_0 = h$, then a and b simplify to $a = 2/h^3$ and $b = \frac{1}{2}h - x_0$, and H_0 is completely determined.

It is easier to construct $K_0(x)$. Here we observe that x_0 is a single zero of K_0 whereas x_1 is a double zero, so

$$K_0(x) = c(x - x_0)(x - x_1)^2,$$

and c is determined from the slope condition at x_0: $K_0'(x_0) = 1$. This gives

$$c(x_0 - x_1)^2 + 2c(x_0 - x_0)(x_0 - x_1) = 1,$$

and so

$$c = (x_0 - x_1)^2 = h^{-2}.$$

The remaining two cardinal functions are obtained in a similar way. Here they are set out in full:

$$H_0(x) = \frac{2}{h^3}\left(x - x_0 + \frac{h}{2}\right)(x - x_1)^2, \qquad H_1(x) = -\frac{2}{h^3}(x - x_0)^2\left(x - x_1 - \frac{h}{2}\right),$$

$$K_0(x) = \frac{1}{h^2}(x - x_0)(x - x_1)^2, \qquad K_1(x) = \frac{1}{h^2}(x - x_0)^2(x - x_1). \qquad (2.9)$$

Recall that for *any* four pieces of data f_0, f_0', f_1, f_1' at x_0 and x_1, the unique Hermite interpolating polynomial from $\mathcal{P}_3$ is now given by substitution in Eq. (2.8). As a simple example, the interpolant determined by $f_0 = f_1 = 1$, $f_0' = f_1' = -1$ is shown in Fig. 2.8.

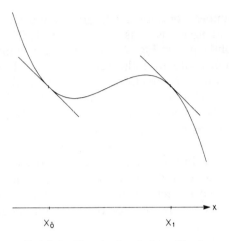

$X_{\dot{0}}$ X_1

FIG. 2.8   Hermite interpolant; $N = 1$.

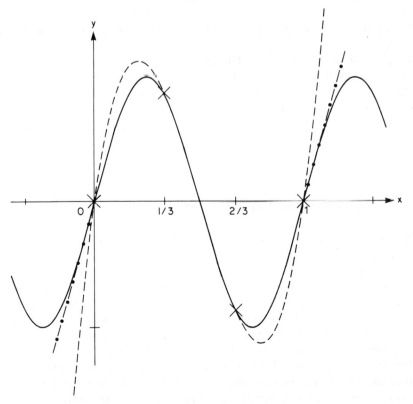

FIG. 2.9   Lagrange and Hermite interpolants to $\sin(2\pi x)$ on four abscissas; $N = 3$.

As a second example, consider the Lagrange and Hermite interpolants to $f(x) = \sin(2\pi x)$ at four abscissas, $x = 0$, $\frac{1}{3}$, $\frac{2}{3}$, 1. The graph of $f(x)$ is indicated by the solid curve in Fig. 2.9, and the four points on the curve at which we are to interpolate are indicated by crosses. The Lagrange interpolant reproduces the four function values only and is in the class of polynomials $\mathscr{P}_3$. The Hermite interpolant reproduces both function and slope values at the four abscissas and is in $\mathscr{P}_7$. The Lagrange interpolant deviates from $f(x)$ by as much as 30% in [0, 1], whereas the error in the Hermite interpolant *on this interval* is not visible on the graph and, in fact does not exceed 0.002. In both cases, note the rapidly increasing error as x increases beyond the extreme abscissa, $x_3 = 1$. In general, if there are $N + 1$ abscissas, we may expect the Hermite interpolant (of degree $2N + 1$) to follow the trend of the data better than the Lagrange interpolant (of degree N).

One way of avoiding high-degree polynomials is to join adjacent pairs of data points with polynomials of some degree, different from point to point, perhaps, and to make sure that where these join, a certain amount of "smoothness" is achieved. Spline functions offer such a possibility and are discussed in Chapter 4. Other forms of piecewise polynomial interpolation are dealt with in Chapter 3.

There is no difficulty in principle in the treatment of polynomial interpolation when the number of consecutive derivatives assigned varies from one abscissa to another [call $f(x_i)$ a zeroth-order derivative at x_i]. [For discussion and implementation of algorithms for this case, see Section 2.7 of Conte and de Boor (1980).]

2.5 Minimizing the sum of squared deviations

The concept of fit of a curve used so far in this chapter has been that of interpolation. We now consider another idea that also qualifies as a method of fitting a curve. The data consists of $N + 1$ distinct abscissas x_0, x_1, x_2, $\ldots$, x_N at which ordinates $f_0, f_1, f_2, \ldots, f_N$ are assigned. The functions we admit in making the fit are those of $\mathscr{P}_m$, where $m \leqslant N$. We consider a typical function $p \in \mathscr{P}_m$ and call $p(x_i) - f_i$ the *deviation of p from the data* at x_i (cf. Fig. 2.10). If a function were to interpolate the data, then each deviation would be zero. Our point of view now is to settle for less and admit non-zero deviations for the sake of using a *simple* function p, i.e. we anticipate that m will be very much less than N. This introduction will be empirical, and we shall return subsequently to some discussion of the philosophy of the approach adopted.

For any $p \in \mathcal{P}_m$, we consider the *sum* of the squared deviations:

$$E(p) = \sum_{i=0}^{N} [p(x_i) - f_i]^2. \tag{2.10}$$

The notation $E(p)$ resembles that used for function values but stands for a somewhat more sophisticated idea. Here p is a function rather than a point of some domain S. In this context, E is called a *functional* and assigns a number to p according to the rule indicated in Eq. (2.10).

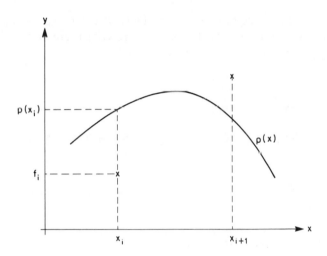

FIG. 2.10   Deviations of $p(x)$ from the data.

Our concept of fit is now prescribed as follows: find a polynomial $p \in \mathcal{P}_m$ for which E is minimized. We shall see that there is only one such minimizing polynomial. This polynomial is, of course, determined by $m + 1$ coefficients $a_0, a_1, \ldots, a_m$. We can view E of Eq. (2.10) as a function of the $m + 1$ coefficients of p and use multivariable calculus to formulate necessary conditions for E to be minimized. This approach to fitting was also introduced by Lagrange, and he named it the *method of least squares*.

The necessary conditions referred to are (Thomas, 1972, p. 522) that the partial derivatives $\partial E/\partial a_i$ vanish for $i = 0, 1, \ldots, m$. (Indeed, this is necessary for any stationary value of E.) Observe that, since E is considered a function of $a_0, \ldots, a_m$, these derivative conditions will provide $m + 1$ equations in these $m + 1$ coefficients. We proceed to formulate these equations. For $j = 0, 1, \ldots, m$, we have [recalling that $p(x) = a_0 + a_1x + \ldots + a_mx^m$]:

$$0 = \frac{\partial E}{\partial a_j} = \sum_{i=0}^{N} 2[p(x_i) - f_i]\frac{\partial p(x_i)}{\partial a_j} = 2\sum_{i=0}^{N} x_i^j[p(x_i) - f_i]$$

$$= 2\left[\sum_{i=0}^{N} x_i^j(a_0 + a_1x_i + \ldots + a_mx_i^m) - \sum_{i=0}^{N} x_i^j f_i\right]$$

$$= 2\left[\left(\sum_{i=0}^{N} x_i^j\right)a_0 + \left(\sum_{i=0}^{N} x_i^{j+1}\right)a_1 + \ldots + \left(\sum_{i=0}^{N} x_i^{j+m}\right)a_m - \left(\sum_{i=0}^{N} x_i^j f_i\right)\right].$$

Observe that these equations are in fact *linear* in $a_0, \ldots, a_m$ and the coefficients are prescribed by the data. We rewrite them as an array (recall that for any real number a, we define $a^0 = 1$):

$$a_0\sum_{i=0}^{N} x_i^0 + a_1\sum_{i=0}^{N} x_i + \ldots + a_m\sum_{i=0}^{N} x_i^m = \sum_{i=0}^{N} f_i,$$

$$a_0\sum_{i=0}^{N} x_i + a_1\sum_{i=0}^{N} x_i^2 + \ldots + a_m\sum_{i=0}^{N} x_i^{m+1} = \sum_{i=0}^{N} x_i f_i, \qquad (2.11)$$

$$a_0\sum_{i=0}^{N} x_i^m + a_1\sum_{i=0}^{N} x_i^{m+1} + \ldots + a_m\sum_{i=0}^{N} x_i^{2m} = \sum_{i=0}^{N} x_i^m f_i.$$

These are known as the *normal equations* for the problem. It is reassuring to know that this algebraic system always has a unique solution and that this solution defines a *minimum* of $E(p)$.

Theorem 2.5.1 *If $x_0, x_1, \ldots, x_N$ are distinct points and $m \leq N$, then there is a unique polynomial $p(x) \in \mathcal{P}_m$ for which $E(p)$ is minimized, and the coefficients of this polynomial are given by the solution of the normal equations (2.11).*

An interesting special case is that in which $m = N$. Here, the number of coefficients equals the number of data points and the minimum value of $E(p)$ is zero. This is because the minimizing polynomial is precisely the *interpolating* polynomial from $\mathcal{P}_N$ for which each deviation in the sum of Eq. (2.10) is zero. The algebraic reason for the coincidence will be indicated shortly. First let us consider a simple example.

EXAMPLE   Fit a straight line ($m = 1$) to the following data by the least squares method.

i	x_i	f_i
0	2	6.91
1	4	9.62
2	6	9.74
3	8	10.01
4	10	11.48
5	12	11.90

Here $N = 5$, $\Sigma x_i = 42$, $\Sigma x_i^2 = 364$, $\Sigma f_i = 59.66$, and $\Sigma x_i f_i = 448.42$. The normal equations are

$$6a_0 + 42a_1 = 59.66,$$

$$42a_0 + 364a_1 = 448.42,$$

giving

$$a_0 = 6.683, \qquad a_1 = 0.4400,$$

and the best least squares straight line is (cf. Fig. 2.11)

$$p(x) = 0.4400x + 6.863.$$

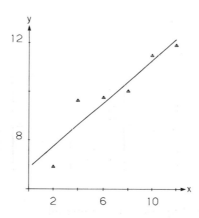

FIG. 2.11   Least squares fit of a straight line
(i.e. a polynomial $p(x)$ from $\mathscr{P}_1$).

Another example is illustrated in Fig. 2.12 for which we omit the computational details. The data are taken from the ramp/mountain function of Fig. 2.4. The abscissas are 17 equally spaced points ($N = 16$). Figure 2.12 indicates two least squares fits, one of sixth degree ($m = 6$) and the other of ninth degree ($m = 9$).

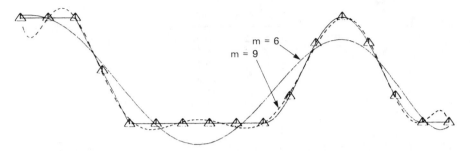

FIG. 2.12   Two least squares fits to the ramp-and-mountain function; $N = 17$.

2.6 Matrix-vector form of the normal equations

It is illuminating, and will subsequently be useful, to describe the normal equations in matrix form. Thus, if X is the $(m + 1) \times (m + 1)$ matrix with elements

$$x_{rs} = \sum_{i=0}^{N} x_i^{r+s}, \qquad r, s = 0, 1, \ldots, m,$$

and we introduce column vectors

$$\mathbf{a} = \begin{bmatrix} a_0 \\ a_1 \\ \cdot \\ \cdot \\ \cdot \\ a_m \end{bmatrix}, \qquad \hat{\mathbf{f}} = \begin{bmatrix} \sum f_i \\ \sum x_i f_i \\ \cdot \\ \cdot \\ \cdot \\ \sum x_i^m f_i \end{bmatrix}, \qquad \mathbf{f} = \begin{bmatrix} f_0 \\ f_1 \\ \cdot \\ \cdot \\ \cdot \\ f_N \end{bmatrix}, \qquad (2.12)$$

then Eq. (2.11) are simply $X\mathbf{a} = \hat{\mathbf{f}}$. Define an $N \times (m + 1)$ matrix V in terms of the abscissas by

$$V = \begin{bmatrix} 1 & x_0 & x_0^2 & \ldots & x_0^m \\ 1 & x_1 & x_1^2 & \ldots & x_1^m \\ & & \vdots & & \\ 1 & x_N & x_N^2 & \ldots & x_N^m \end{bmatrix}. \qquad (2.13)$$

Then it is easily verified, by using the definition of matrix multiplication, that $X = V^T V$ and $\hat{\mathbf{f}} = V^T \mathbf{f}$. Thus, the normal equations $X\mathbf{a} = \hat{\mathbf{f}}$ can also be written in the factored form:

$$V^T V \mathbf{a} = V^T \mathbf{f}. \tag{2.14}$$

It is no accident that the matrix V is a more general form of the Vandermonde matrix introduced in Eq. (2.2) in connection with polynomial interpolation. Note, in particular, that if $m = N$ then V is square and (as noted in Section 2.2) has an inverse. It follows that $(V^T)^{-1}$ also exists in this case, and so multiplying Eq. (2.14) on the left by $(V^T)^{-1}$, we obtain $V\mathbf{a} = \mathbf{f}$ and then $\mathbf{a} = V^{-1}\mathbf{f}$, the solution of the polynomial interpolation problem. Thus, as remarked above, the least squares fit is just the polynomial interpolant in this special case.

The warning remarks made in Section 2.2 concerning direct computation with V are doubly appropriate when applied to the normal equations. The factorization $X = V^T V$ of the coefficient matrix can lead to fiercely ill-conditioned numerical problems. It is necessary to design general-purpose computer codes for this problem with great care. The interested reader is referred to Lawson and Hanson (1974) or chapter 7 of Forsythe *et al.* (1977) for detailed discussion and codes.

2.7 The rationale for least squares methods: polynomial regression

Our introduction of the least squares process has been empirical—as a process which, by minimizing the sum of squared deviations, provides a function that may represent a mean value (or even the value) of a function from which the data has been sampled. There are other ways of finding such a mean function, however, which may look quite different from the least squares fit. For example, if we continue working with the deviations, we could equally well choose to find $p(x) \in \mathcal{P}_m$ so as to minimize

$$\sum_{i=0}^{N} |p(x_i) - f_i|, \quad \text{or} \quad \max_{0 \leq i \leq N} |p(x_i) - f_i|,$$

instead of $E(p)$ of Eq. (2.10).

The primary reason for choosing $E(p)$ is simply computational convenience. In the absence of a good reason for choosing another of these criteria, one would choose $E(p)$ because the resulting least squares fit is much the easiest to compute. The other two criteria suggested here have been the subjects of analysis, but they give rise to *nonlinear* equations for the coefficients of p, requiring rather more mathematical expertise for their solution than the confines of this volume allow us to present.

There are, however, certain problems where the least squares technique has a rational justification. The relevant process is of a statistical nature and is known as *polynomial regression*. To give a quick introduction to this we first need the notion of a normal probability distribution. Here, it is supposed that the outcome of a statistical experiment can take any real numerical value. A *probability distribution* is then a function $n(x)$ defined for $x \in (-\infty, \infty)$ with the property that the probability that a measured outcome y of the experiment is between α and β (for any real α and β) is (where P is probability)

$$P(\alpha < y < \beta) = \int_\alpha^\beta n(x)\, dx.$$

Since certainty is measured by a probability 1 and all probabilities are non-negative, a probability distribution must satisfy the conditions

$$\int_{-\infty}^{\infty} n(x)\, dx = 1, \tag{2.15}$$

and $n(x) \geq 0$ for all real x.

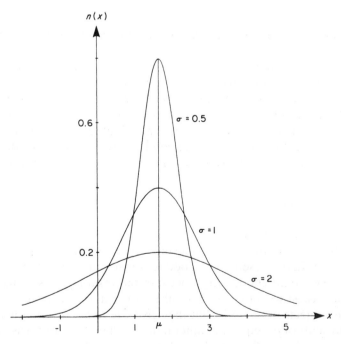

FIG. 2.13   Three normal distributions.

The *normal* probability distribution with mean μ and variance σ^2 is defined by the function

$$n(x) = (\sqrt{2\pi}\sigma)^{-1} \exp\{-(x - \mu)^2/2\sigma^2\}. \qquad (2.16)$$

Three such functions are plotted in Fig. 2.13. Changing μ simply shifts the curve to the right or left without changing its shape. Thus, the distributions illustrated have the same mean with different variances.

The verification of property (2.15) for the normal distribution is a mathematical nicety that the reader will find in books on the theory of statistics or of probability.

The point of view of regression theory is to interpret a piece of data (x_i, f_i) as follows: x_i is a (precisely known) value of an independent variable or parameter, and f_i is the outcome of an experiment conducted at $x = x_i$ in which the probability distribution is normal. The mean of this distribution is supposed to be a function of the independent variable x, whereas the variance is *constant*. Polynomial regression is then a process that requires that μ be a *polynomial* function of x, say,

$$\mu(x) = \sum_{i=0}^{m} a_i x^i \in \mathcal{P}_m, \qquad (2.17)$$

so that the probability distribution at x has the form

$$n_x(y) = (\sqrt{2\pi}\sigma)^{-1} \exp\{-(y - \mu(x))^2/2\sigma^2\}. \qquad (2.18)$$

The situation is illustrated in Fig. 2.14 in which $y = \mu(x)$ is the *regression curve*.

The primary purpose of regression theory is then to obtain estimates of the coefficients $a_0, \ldots, a_m$ in Eq. (2.17), which determine $\mu(x)$, and of the variance σ^2. This is done by observing that the probability that f_i is observed at x_i, where $i = 0, 1, 2, \ldots, N$ is

$$\prod_{i=0}^{N} n_{x_i}(f_i) = \frac{1}{(2\pi)^{(N+1)/2} \sigma^{N+1}} \exp\left\{-\sum_{i=0}^{N} (f_i - \mu(x_i))^2/2\sigma^2\right\}.$$

Then an estimate of the regression curve is determined by that choice of function $\mu(x) \in \mathcal{P}_m$ which maximizes this probability of the $N + 1$ actual observations. This is seen to be equivalent to minimizing the exponent on the right, and this in turn is equivalent to minimizing $E(\mu)$ [cf. Eq. (2.10)]. Thus, the estimated regression curve $\mu(x)$ is just the polynomial in $\mathcal{P}_m$ determined by the method of least squares.

Note that this process determines *estimates* of the coefficients of the regression curve and of σ^2, which themselves have associated probability distributions. Following up this observation leads to an "unbiased" estimate of σ^2:

$$\hat{\sigma}^2 = \frac{1}{N-m} \sum_{i=0}^{N} (f_i - \hat{\mu}(x_i))^2,$$

where $\hat{\mu}(x) = \sum_{i=0}^{m} \hat{a}_i x^i$ and $\hat{a}_0, \ldots, \hat{a}_m$ are given by the solution of the normal equations. The summation in $\hat{\sigma}^2$ is, not unnaturally, called the residual sum of squares.

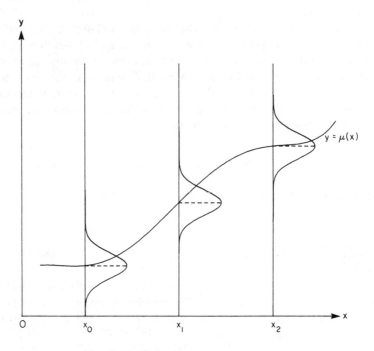

FIG. 2.14   Model for polynomial regression.

The polynomial $\hat{\mu}(x)$ has two convenient properties, namely,

$$\sum_{i=0}^{N} \hat{\mu}(x_i) = \sum_{i=0}^{N} f_i$$

and

$$\sum_{i=0}^{N} \hat{\mu}(x_i)^2 = \sum_{i=0}^{N} f_i \hat{\mu}(x_i).$$

(These can be confirmed using equation (2.14) and the fact that $[\hat{\mu}(x_0), \ldots, \hat{\mu}(x_N)] = \hat{a}^T V^T$.) The first implies that the mean of the values of the fitted curve at the data points is equal to the mean of the values f_0, $f_1, \ldots, f_N$ at the same points. Using the second of these, it is not difficult to see that $\hat{\sigma}^2$ can be written in a more easily computed form:

$$\hat{\sigma}^2 = \frac{1}{N - m} \left(\sum_{i=0}^{N} f_i^2 - \sum_{i=0}^{N} f_i \hat{\mu}(x_i) \right). \tag{2.19}$$

The advantage of this approach, when it applies, is that there is a more objective theory behind it than the simple least squares argument. This theory admits statistical evaluation of the goodness of the fit and confidence intervals for the coefficients of $\mu(x)$. In numerical practice, when the best choice of m is not clear from the data or the underlying phenomenon, a value of $\hat{\sigma}^2$ is obtained from Eq. (2.19) for a few integers m, and the smallest m making $\hat{\sigma}^2$ reasonably small may be chosen as most appropriate. This choice makes the distribution curves on Fig. 2.14 narrower and more peaked. More sophisticated tests are available based on significance tests of the hypothesis that some coefficients $a_k, \ldots, a_m$ are zero. A useful survey is given by Hemmerle (1967). For the statistical theory, we refer the reader to Mood and Graybill (1963).

To conclude this account of the (polynomial) regression approach, let us review once more the usual working assumptions that are made: the ordinates f_i are subject to random error; with each x there is an associated probability distribution (usually the *normal* distribution) with mean μ depending on x and variance σ^2 independent of x; there is *a priori* information allowing one to say that, for some m (usually as small as possible), $\mu(x) \in \mathcal{P}_m$.

2.8 Smoothing, or finding a trend?

The preceding discussion of regression introduces a notion of statistical variation into the data. This may be appropriate in some physical situations and not in others. This point serves to illustrate the distinction between smoothing of the data and finding a trend in the data, both of which are widely used ideas.

The point of view of smoothing is that the data is subject to error, but that there is some underlying relationship between abscissas and ordinates that would have certain properties not apparent in the perturbed data. For example, there may be good reason to believe that there is an underlying function that has slope varying only slowly in the x-domain of interest. An *interpolant* to the perturbed data may, however, have a wildly fluctuating

slope and therefore be unacceptable as a model for the underlying phenomenon. To get an acceptable representation of slopes, it may then be
necessary to smooth the data. The smoothing process itself may simply be
empirical or, if there is a theory (such as regression) whose hypotheses are
satisfied by the phenomenon being modelled and which is computationally
acceptable, then this would likely be preferred.

The process of smoothing is sometimes referred to as *graduation*, and is
perhaps best known under this name in actuarial work. A great deal of this
work uses the hypothesis that the abscissas $x_0, x_1, \ldots, x_N$ are equally
spaced. This is an assumption that we do not care to make at this point.
Furthermore, several classical methods can now be superseded by moving
least squares techniques and by spline function methods, both of which will
be discussed in the sequel. Consequently, we shall not linger on these
topics, but refer the reader to Chapter 11 of the classic work by Whittaker
and Robinson (1944) or to the more modern presentation of Davis (1973).
There are also points of contact here with methods of moving averages
used in "filtering" techniques. We shall omit specific discussion of these
for similar reasons.

The point of view of the analysis of trends is rather different, although
the process for numerical implementation (via the normal equations, for
example) could be exactly the same as for smoothing. In this case, the data
need not be seen as being subject to error, but there is some reason for
seeing the ordinates as the sum of two components. There may be a good
physical reason for this, each component being the result of a different
physical process, probably on different scales [of the independent
variable(s), x in this case]. The component varying more slowly with the
independent variable is then conveniently described as the trend and the
other as the residual or deviation from the trend. In the parlance of
geophysics, the trend may be known as the "regional" component.

On the other hand, this separation into two components may be simply
for convenience. For example, the detection and removal of a trend may
facilitate computation with a residual that is thought to contain all the
information of interest. From the point of view of frequency analysis, one
may attempt to remove undesirable low-frequency components in this way.

Whether the appropriate context is smoothing or trend analysis, the
following quantity is frequently used empirically as a measure of the degree
to which the trend, of a smoothed curve, fits the data. Define

$$R^2 = 1 - \frac{\Sigma (f_i - \hat{\mu}(x_i))^2}{\Sigma (f_i - \bar{f})^2} = \frac{\Sigma (\bar{f} - \hat{\mu}(x_i))^2}{\Sigma (f_i - \bar{f})^2}, \qquad (2.20)$$

where $\bar{f} = (\Sigma f_i)/(N + 1)$ is just the arithmetic mean of the $N + 1$ data values. The expression in the denominator of these fractions is then just the variance of all the observations about this mean, often written in the alternative form $(\Sigma f_i^2) - (\Sigma f_i)^2/(N + 1)$.

Note that $0 \leqslant R^2 \leqslant 1$, and when $R = 1$ we have $\hat{\mu}(x_i) = f_i$ for each i so that all of the data lies on the smoothed, or regression, curve; i.e. this curve *interpolates* the data, and there is no residual or deviation. On the other hand, when $R = 0$, $\hat{\mu}(x_i) = \bar{f}$ (a constant) for each i, and the analysis has, in effect, failed to show any trend. Thus, values of R^2 between 0.8 and 1 are often thought to show a marked or significant trend in the data, and values of R^2 between 0 and 0.2 suggest that the trend is not well established.

The parameter R^2 is also significant from the point of view of statistical theory, but this is in the context of maximum likelihood estimation rather than the regression theory introduced above. This means, of course, a different theory and a different set of statistical hypotheses to satisfy [See Hemmerle (1967) for accessible comments on statistical theory and Unwin (1975) for the point of view of a geographer on trend surface analysis.]

2.9 Moving least squares

In this section we introduce a variant of the method of least squares. What we describe here is not recommended as a viable fitting procedure in its own right. The discussion is purely expository and leads to interesting interpolation procedures whose descriptions are completed next in Section 2.10. The "weighted least squares" method to be presented here and the "interpolating moving least squares" method of Section 2.10 find their main application in (multivariate) surface interpolation with scattered data and are discussed in that context in Chapter 10.

We first modify the point of view adopted in Section 2.5 for the method of least squares and propose that, if g is to be the function associated with the fitted curve, then the value of g at a point x should be most strongly influenced by the data (the numbers f_i) at those points x_i that are closest to x. For the same reason, it is proposed that as the distance of x_i from x increases, the influence of the data on g at x should decrease. This can be accomplished very easily if, instead of minimizing a sum of squared deviations, as formulated in Eq. (2.10), we associate positive weights w_i (which will depend on x) with each deviation and minimize

$$E_x(p) = \sum_{i=0}^{N} w_i(x)[p(x_i) - f_i]^2. \tag{2.21}$$

For the time being, we assume that the numbers $w_i(x)$ are to be chosen so that they are positive, relatively large for the x_i close to x, and relatively small for the more distant x_i. More precisely, we assume that $w_i(x)$ decreases monotonically as $|x - x_i|$ increases. As in Section 2.5, we assume here that p is a polynomial in $\mathscr{P}_m$, say $p(x) = \sum_{i=0}^m a_i x^i$.

The normal equations are found just as before by consideration of the $m + 1$ necessary conditions $\partial E_x / \partial a_i = 0$, $i = 0, 1, \ldots, m$. The resulting normal equations for the coefficients $a_0, a_1, \ldots, a_m$ look very much like Eq. (2.11). The latter are modified by insertion of the weights under the summation signs as follows:

$$
\begin{aligned}
(\Sigma\, w_i x_i^0) a_0 + \ldots + (\Sigma\, w_i x_i^m) a_m &= \Sigma\, w_i f_i, \\
(\Sigma\, w_i x_i) a_0 + \ldots + (\Sigma\, w_i x_i^{m+1}) a_m &= \Sigma\, w_i x_i f_i, \\
&\vdots \\
(\Sigma\, w_i x_i^m) a_0 + \ldots + (\Sigma\, w_i x_i^{2m}) a_m &= \Sigma\, w_i x_i^m f_i.
\end{aligned}
\tag{2.22}
$$

In Eq. (2.22), we have abbreviated $w_i(x)$ as w_i, and as long as all of these numbers are positive, a theorem such as Theorem 2.5.1 obtains. That is, we are guaranteed a unique solution for $a_0, a_1, \ldots, a_m$, say $\hat{a}_0, \ldots, \hat{a}_m$, and the polynomial $\hat{p}$ that they define does indeed give the least value to E_x of all polynomials in $\mathscr{P}_m$. Note also that Eq. (2.11) can be obtained from (2.22) by the choice of weights $w_i(x) = 1$, $i = 0, 1, 2, \ldots, N$.

Since the coefficients of (2.22) depend on x (through the w_i), the solution $\hat{a}_0, \ldots, \hat{a}_m$ will also depend on x. For the ordinate of the function g at x (which we are now constructing), we take

$$
g(x) = \hat{p}(x). \tag{2.23}
$$

Now this prescription for the value of $g(x)$ applies for any x whatever. Thus, we have given a constructive definition of the function g, defined on the whole real line. The snag is, that to evaluate g we have to be prepared to expend the effort of solving a set of normal equations *for each x*. For this reason, the process is generally only applied for small values of m, say, 0, 1, or 2. This function g may be described as a *moving least squares* fit to the data.

It is important to keep in mind that, in spite of the relation (2.23), valid for fixed x, g will not generally be a polynomial function. This is because the coefficients of $\hat{p}$ in Eq. (2.23) also depend on x. To illustrate the point, we consider the data of Fig. 2.11 again. As with the least squares fit in that illustration we take $m = 1$. To calculate a *moving* least squares fit we have to specify a distribution of weights for each x.

Consider the function e^{-x^2}, also written as $\exp(-x^2)$. For $x \geqslant 0$, it is a monotonically decreasing function of x. In view of the abscissas of our example, it is convenient to consider the function $\exp\{-x^2/50\}$ for $x \geq 0$ (shown in Fig. 2.15) and then to define

$$w_i(x) = \exp\{-(x - x_i)^2/50\}, \tag{2.24}$$

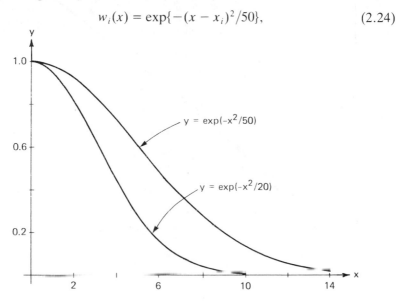

FIG. 2.15   Two exponential functions.

In this way we use the same underlying exponential function for each i; only its centre of symmetry depends on i. [Note that this ubiquitous function has already appeared in Eq. (2.16).] The resulting moving least squares fit f is illustrated in Fig. 2.16.

What desirable properties does a function g constructed by the moving least square process possess? First, it has a *reproducing* property (which it shares with the least squares fit). Namely, if the data all happen to be on the graph of a function from $\mathcal{P}_m$ (and polynomials $\hat{p}$ from $\mathcal{P}_m$ are used in the construction of g), then g will be this same polynomial from $\mathcal{P}_m$. Second, the function g is smooth in the sense that it can be differentiated repeatedly (for as many times as w is differentiable). Thus, we have $g \in \mathscr{C}^\infty$.

To complete this preparatory section, it will also be useful to examine the matrix-vector form of Eqs. (2.22) in anticipation of a generalization of Eq. (2.14). In addition to the definitions of Eqs. (2.12) and (2.13), we introduce an $(N + 1) \times (N + 1)$ diagonal matrix:

$$W(x) = \mathrm{diag}[w_0(x), \quad w_1(x), \quad w_2(x), \quad \ldots, \quad w_N(x)]. \tag{2.25}$$

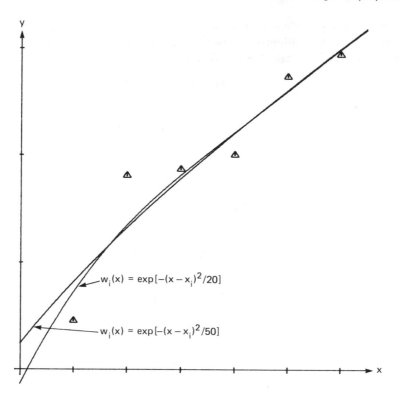

$$w_i(x) = \exp[-(x - x_i)^2/20]$$

$$w_i(x) = \exp[-(x - x_i)^2/50]$$

FIG. 2.16   Two moving least squares fits (data of Fig. 2.11; $M = 1$).

The matrix product $V^T W V$ is then $(m + 1) \times (m + 1)$, and it is easily verified that, in matrix form, the normal equations [Eqs. (2.22)] are, more concisely,

$$V^T W(x) V \mathbf{a} = V^T W(x) \mathbf{f}. \tag{2.26}$$

This obviously reduces to Eq. (2.14) when $W(x)$ is the identity matrix (and so is independent of x).

2.10 Interpolating moving least squares (IMLS) methods

Comparing Figs. 2.11 and 2.16 to contrast the least squares and moving least squares fits, we observe that the effect of introducing variable weights has been to draw the fitted curve towards each data point. The idea behind the interpolation process is to assign weights in such a way that this trend towards interpolation of the data is carried to its limit. This is achieved by readjustment of the relative weights.

To see the effect of such a readjustment consider the function $\exp\{-x^2/20\}$, also illustrated in Fig. 2.15, which is more peaked than $\exp\{-x^2/50\}$. If we assign weights, as before, by putting

$$w_i(x) = \exp\{-(x - x_i)^2/20\},$$

this will put *relatively* higher weights on points x_i close to x and lower weights on points far removed from x. The resulting fitted curve shown in Fig. 2.16 comes closer to interpolating the data then that obtained with more evenly distributed weights. Note that, because the weights appear in a similar way on both sides of Eq. (2.22) or Eq. (2.24), the w_i's can be scaled (i.e. all multiplied by the same positive number) without affecting the solution of the equations and hence the fitted curve defined by g.

To go to the extreme and ensure interpolation at x_i, the trick is to use a weight function $w_i(x)$ that becomes infinite at x_i, a property which will also be invariant under scaling. By assigning different weight functions to different points, some with vertical asymptotes and some without, we can even arrange for interpolation at some points and not at others.

For simplicity, we shall always assume [as in Eq. (2.24)] that a simple function $w(x)$ is prescribed that is monotonically decreasing for $x > 0$ and then assign weight functions $w_i(x)$ by defining

$$w_i(x) = w(|x - x_i|), \qquad i = 0, 1, 2, \ldots, N. \tag{2.27}$$

In this way, if $w(x) \to \infty$ as $x \to 0$, then we will have $w_i(x) \to \infty$ as $x \to x_i$ for *every i*.

From the point of view of analysis, one might anticipate difficulties in the solution of Eqs. (2.22) and (2.26) near a data point x_i, because the coefficients "blow up" near x_i. However, this does not necessarily lead to difficulties, largely as a result of the homogeneity in the weights already referred to. Indeed, it can be proved that under quite general assumptions about the weight function and in spite of singularities in the weights, the equations have a unique solution for every x, that these solutions define a curve given by $y = g(x)$, and that the function g retains the reproducing and smoothness properties described in Section 2.9 for the non-interpolating case [cf. Lancaster and Šalkauskas (1981)].

For interpolating moving least squares methods, the exponential functions used previously in Section 2.9 are no longer candidates for weight functions since they do not have the right asymptotic behaviour as $|x| \to 0$. The most popular choices probably involve inverse even powers of $|x|$. In particular, if $w(x) = 1/x^2$ or $1/x^4$, we would obtain either

$$w_i(x) = 1/(x - x_i)^2 \qquad \text{or} \qquad w_i(x) = 1/(x - x_i)^4.$$

Alternatively, $w(x) = \exp(-x^2)/x^2$ would have similar properties to $w(x) = 1/x^2$ near $x = 0$ but would attenuate more rapidly as $|x|$ increases. In general, the user of the technique must decide on the proper weight function on such qualitative grounds as the rate of attenuation in relation to density of abscissas, the need for interpolation, and the nature of the singularity at $x = 0$. Fortunately, these are the only important questions since in general the shape of the fitted curve is not sensitive to the precise nature of the chosen weight function.

More generally, considering inverse distance weight functions of the form $w(x) = x^{-k}$, $k > 0$, it is found that the *smoothness* of g (in the sense of differentiability) depends critically on k. We shall return to this point in Section 2.11 where we discuss Shepard's method. For the time being, we assume that g is always to be differentiable, and for this to be the case it is necessary that $k > 1$. In fact, this should be seen as a necessary condition on the nature of the singularity introduced into the weight function to ensure differentiability of g at the data points. If the function g is to be $\mathscr{C}^\infty$, it is necessary that k be a positive, even integer.

For example, the formula

$$w(x) = \alpha \exp(-\beta x^2)/x^k, \tag{2.28}$$

defines a family of weight functions dependent on three positive parameters α, β, and k. Since $\exp(-\beta x^2)$ is a function with horizontal tangent at $x = 0$ (cf. Fig. 2.15), the nature of the singularity in w at $x = 0$ is determined by the exponent k. As with a simple inverse power law (when $\beta = 0$), differentiability of g will be retained as long as $k > 1$. The main disadvantage of a choice such as Eq. (2.28) with $\beta > 0$ is the added computational expense implied by many evaluations of the exponential function.

Before going on to discuss examples, there is one other freedom in the choice of weight function that is very important to applications. The idea here is to consider the extreme case of attenuation of the weight function for large x. Instead of arranging for w to become very small for large $|x|$, we arrange for w to be *zero* for all sufficiently large x. For interpolation we would like to retain the infinity in w at $x = 0$, and for cosmetic reasons, we would like w to be at least once differentiable for all $x \neq 0$. A function w that satisfies these three desiderata is, for example,

$$w(x) = \begin{cases} ax^{-k}\{1 - |x|/d\}^2, & \text{for} \quad |x| \leq d, \\ 0, & \text{for} \quad |x| > d, \end{cases} \tag{2.29}$$

Note that w involves three parameters, a, d, and k, all assumed to be positive, and $w(x) = 0$ for $|x| > d$. As indicated earlier, it would generally be the case that k is a positive, even integer. The function is said to have *support* $(-d, d)$, i.e. w is non-zero for any x in this set and is zero otherwise, and the support is assigned when we fix d. The parameter a is convenient and simply allows us to scale w, having no effect on the final fitted curve. Two such weight functions are illustrated in Fig. 2.17. The function $(1 - |x|/d)^2$ is also indicated for comparison.

Similar truncated weight functions are described by the family

$$w(x) = \begin{cases} ax^{-k} \cos^2(\pi x/2d), & \text{for} \quad |x| \le d, \\ 0, & \text{for} \quad |x| > d, \end{cases} \tag{2.30}$$

although these will be rather more expensive in computer time due to the presence of the cosine function.

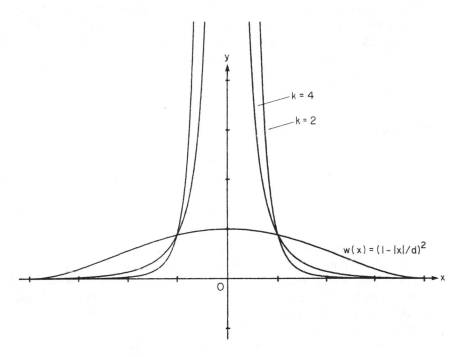

FIG. 2.17 Two truncated weight functions.

Careful adjustment of the diameter of support for the weight function (i.e. of the parameter d) is most important. It must be large enough so that, for every x at which g is to be defined, the interval $(x - d, x + d)$ contains at least $m + 1$ data points. At an x where this is not satisfied, the normal equations will have a singular coefficient matrix and the process will break down.

In terms of the matrices V and $W(x)$ of Eq. (2.24), we must have $m + 1$ non-vanishing weights $w_i(x)$ appearing on the diagonal of W. In this case, the $(m + 1) \times (m + 1)$ matrix $V^T W(x) V$ will be invertible, and the moving least squares calculations can be completed.

There are two principal advantages in truncation of the weight function. The first has to do with economy of computation. In formulating the coefficients of the normal equations [in the form of Eq. (2.22)], the summations can be restricted to only those i for which w_i is non-zero. It must be noted, however, that this can only be done at the expense of careful programming. We must have a good procedure to decide for which values of i it is true that $|x - x_i| < d$.

The second advantage is qualitative and can be stated concisely by saying that it produces a *local* interpolation scheme. This means that the value of g (determining the fitted curve) at x is determined only by the data sufficiently close to x. For example, we can be quite sure that if the weight function at x has support $(x - d, x + d)$, then perturbations in the data outside this interval will not affect $g(x)$.

2.11 Examples of the IMLS methods

(*a*) *The case* $m = 0$. *Shepard's scheme.* When $m = 0$, the normal equations of Eq. (2.22) reduce to a single equation:

$$a_0(x) = \sum_{i=0}^{N} w_i(x) f_i \bigg/ \sum_{i=0}^{N} w_i(x), \qquad (2.31)$$

so that the computational effort in the solution of the normal equations is no longer so important. The family of methods in which the weights are determined by inverse powers of the distance, i.e. $w(x) = x^{-k}$ with $k > 0$ together with Eq. (2.27), are associated with the name of Shepard (1968). The differentiability of the interpolating function is relatively easy to determine in this case.

It can be shown that for $0 < k < 1$, the interpolating curve has cusps at the data points (being otherwise $\mathscr{C}^\infty$), and if $k = 1$, the fitted curve has corners at the data points. Thus, we have $g(x) = f_i$ for each i, but g is not differentiable at x_i as long as $0 < k \le 1$. For $k > 1$, g is at least once differentiable everywhere, i.e. $g \in \mathscr{C}^1$. These facts seem to argue in favour of even integers for k where the choice of k is open and, for simplicity, $k = 2$ is to be preferred. Some examples (particularly with $k < 1$) together with analysis are to be found in the paper of Gordon and Wixom (1978); see also the survey of Barnhill (1977).

Before looking at some examples of Shepard's method, note some simple special cases of Eq. (2.31): First, if $w_i(x) = 1$ for $i = 0, 1, 2, \ldots, N$ and all x, then Eq. (2.31) reduces to $a_0 = (\Sigma f_i)/(N + 1)$ which is the average, or arithmetic mean of the ordinates, and the fit is simply a straight line parallel to the x-axis. If the w_i are different but independent of x, a_0 is also independent of x and is a *weighted average* of the ordinates.

A *moving average* is obtained if we choose, for example,

$$w_i(x) = \begin{cases} 1 & \text{if} \quad x_i \in (x - \alpha, x + \alpha], \\ 0 & \text{if} \quad x_i \notin (x - \alpha, x + \alpha], \end{cases}$$

where α is some fixed positive number. In this case $a_0(x)$ does depend on x and is a piecewise constant (or *step*) function. The result of applying this process to the data of Fig. 2.11 with $\alpha = 3$ is illustrated in Fig. 2.18. The weight function in this case is *truncated*, as in Eq. (2.29) or Eq. (2.30), but without the smoothness properties of those examples at the frontier of the

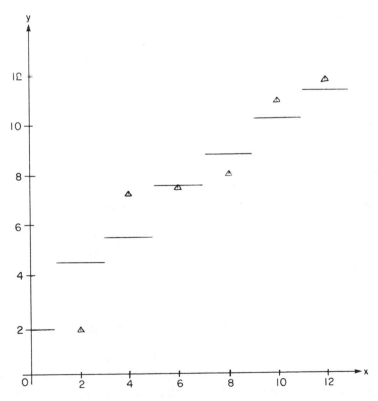

FIG. 2.18   Moving average to the data of Fig. 2.11.

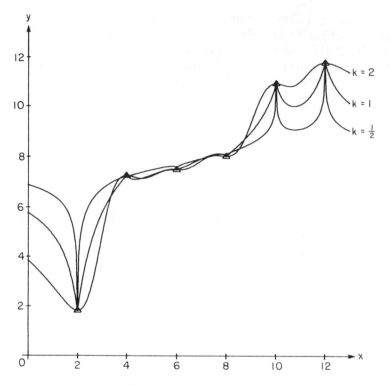

FIG. 2.19   Shepard interpolants with $w(x) = x^{-k}$, $k = \frac{1}{2}, 1, 2$.

support [and hence no smoothness in $a_0(x)$] and without the singularity at $x = x_i$ (and hence no interpolation, in general).

Figure 2.19 shows three Shepard interpolants to the data of Fig. 2.11. The interpolant corresponding to $k = 2$ is at least once differentiable, it exhibits the "flat-spot" phenomenon: The derivative of the interpolant is zero at every data point. This is true for every Shepard interpolant with $k > 1$ and severely limits its usefulness.

(b)   *The case m = 1.*   Now a *pair* of normal equations stemming from Eq. (2.22) must be solved for *each* abscissa x of the interpolant. They are

$$(\Sigma\, w_i(x))a_0 + (\Sigma\, w_i(x)x_i)a_i \quad\ = \Sigma\, w_i(x)f_i,$$

$$(\Sigma\, w_i(x)x_i)a_0 + (\Sigma\, w_i(x)x_i^2)a_1 = \Sigma\, w_i(x)x_i f_i.$$

The computational effort is now bigger. As in the Shepard interpolant, care must be exercised when the calculations are carried out near a data point in view of the singularity of $w(x)$ there. The interpolating curve resulting from the weight $w(x) = x^{-2}$ is shown in Fig. 2.20. The flat-spot phenomenon is no longer present.

(c) *The case m = 2.*   Here, a quadratic fit involving three normal equations is carried out for each point of the graph. For the data of Fig. 2.20, and with the same weight function as in (b), namely $w(x) = x^{-2}$, the resulting interpolant is also shown in Fig. 2.20. There is not much change from the curve using $m = 1$.

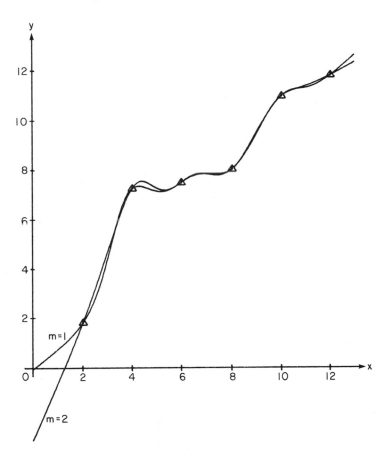

FIG. 2.20   Interpolating moving least squares fits with $N = 1, 2$.

3

Interpolation with Piecewise Polynomial Functions

3.1 Introduction

From the preceding discussion of polynomial interpolation, it is evident that as the number of interpolation points increases, the necessary flexibility of the interpolating polynomial is obtained by increasing both the degree and the risk of severe oscillations of the fitted curve.

In the following sections we shall consider alternatives that allow the degree of the polynomials involved to be kept low and provide flexibility by making use of a sufficient number of polynomial segments joined in a smooth way. The resulting interpolating or smoothing function is called a *piecewise polynomial function*. Perhaps the most famous of these is the *cubic spline*, which is introduced in this chapter but discussed in more detail in Chapter 4. Before proceeding to these rather sophisticated functions, we shall consider the more simple situation involving *piecewise linear* functions, or *linear* splines, which lack smoothness but serve to prepare the ground for our later discussion as well as being useful themselves in some cases.

3.2 Piecewise linear functions: linear splines

In Section 1.7 we introduced the family of functions $\mathcal{P}_N$ (a vector space) consisting of polynomials of degree less than or equal to N. There they were defined as linear combinations of the $N + 1$ very simple functions $1, x, x^2, \ldots, x^N$ [cf. Eq. (1.6)]. Subsequently, either an interpolant or smoothing function was chosen from this family. Here we shall proceed in a similar manner, although a few more complications are necessary. First, we shall insist that the piecewise linear functions that we use be continuous so that their graphs are basically zigzags. Thus, the location of the corners is relevant; we shall call their abscissas *knots*, and they will of course be distinct. In describing graphs of this kind mathematically, the translates of the absolute-value function of Fig. 1.3 can play a useful role. The graph of such a function, $|x - k|$ (there is a simple knot at k) is illustrated in Fig. 3.1. We shall take advantage of such functions in the following definition.

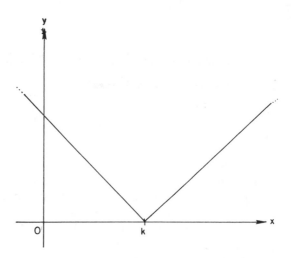

FIG. 3.1   Translate of the absolute-value function.

Definition   *Let K denote the set $\{k_0, \ldots, k_N\}$ of knots (real numbers) satisfying $k_0 < k_1 < \ldots < k_N$. A function l defined for all real numbers x by*

$$l(x) = a_0|x - k_0| + a_1|x - k_1| + \ldots + a_N|x - k_N|,$$

where $a_0, a_1, \ldots, a_N$ are fixed real numbers, is a (continuous) piecewise linear function (or linear spline) with knot sequence K.

In order to see that l is a function of the desired kind, we observe that $|x - k_i|, i = 0, 1, \ldots, N$, is the absolute value function of Fig. 1.3, translated so as to have its corner at $x = k_i$. Observe that l is a linear combination of the continuous functions $|x - k_0|, \ldots, |x - k_N|$ and as a result l is itself continuous. Between any two knots, each of the functions $|x - k_i|, i = 0, \ldots, N$, contributes a straight-line segment to the graph, and thus, l has a graph consisting of straight-line segments joined at the knots. It will have a continuous first derivative for *all* $x \in [a, b]$ only if $a_0 = a_1 = \ldots a_N = 0$. Following the discussion in Section 1.7, we find:

Theorem 3.2.1   *The set of all linear splines l with fixed knot sequence K containing $N + 1$ knots is a vector space of dimension $N + 1$.*

We denote the space of piecewise linear functions of Theorem 3.2.1 by $\mathcal{L}_N(K)$. Observe that every function of $\mathcal{L}_N(K)$ is in the class $\mathscr{C}[a, b]$ introduced in Section 1.3, and so $\mathcal{L}_N(K)$ is known as a *subspace* of $\mathscr{C}[a, b]$.

In our further work we assume that the required interpolation or smoothing is carried out on a finite interval $[a, b]$. Flexibility outside $[a, b]$ is of no interest; however, in order to avoid certain technical difficulties, it will be assumed that $k_0 = a$ and $k_N = b$ so that all the knots of K are in $[a, b]$.

3.3 Interpolation with linear splines

As in Section 2.1, we are required to find a function, this time from $\mathcal{L}_N(K)$, that interpolates the data (x_i, f_i), $i = 0, 1, \ldots, N$, the abscissas x_i being distinct points in $[a, b]$. This is particularly easy if the points x_i are simply the knots k_i, which determine the space $\mathcal{L}_N(K)$. In this case we have Theorem 3.3.1.

Theorem 3.3.1   *Let a knot sequence $K = \{x_0, x_1, \ldots, x_N\}$ and arbitrary real numbers $f_0, f_1, \ldots, f_N$ be given. Then, there is a unique spline l in $\mathcal{L}_N(K)$ that satisfies $l(x_j) = f_j$ for $j = 0, 1, \ldots, N$.*

The nature of this interpolant is of course quite obvious: One connects the data points from left to right by a broken-line segment, bearing in mind that we have adopted the convention $a = x_0 < x_1 < \ldots < x_N = b$. On the other hand, if the number of knots exceeds $N + 1$, or if they are unfortunately located with respect to the x_i's, a unique (or indeed any) interpolant may fail to exist.

In order to determine the coefficients a_i, $i = 0, \ldots, N$ of the interpolant of Theorem 3.3.1 we may proceed as in Section 2.2 and find that in view of the interpolation conditions,

$$a_0|x_0 - x_0| + \ldots + a_N|x_0 - x_N| = f_0,$$
$$\vdots \qquad\qquad\qquad \vdots$$
$$a_0|x_N - x_0| + \ldots + a_N|x_N - x_N| = f_N,$$

and thus in matrix form,

$$
\begin{bmatrix}
0 & |x_0 - x_1| & \ldots & |x_0 - x_N| \\
|x_1 - x_0| & 0 & & \vdots \\
\vdots & & \ddots & |x_{N-1} - x_N| \\
|x_N - x_0| & |x_N - x_1| & \ldots & 0
\end{bmatrix}
\begin{bmatrix}
a_0 \\
\vdots \\
\\
a_N
\end{bmatrix}
=
\begin{bmatrix}
f_0 \\
\vdots \\
\\
f_N
\end{bmatrix},
\tag{3.1}
$$

the matrix being invertible (non-singular).

A much more convenient representation results if a *cardinal basis* is chosen at the outset. This is a set of functions l_i from $\mathscr{L}_N(K)$ such that each l_i solves the elementary interpolation problem

$$l_i(x_k) = \delta_{ik}, \qquad k = 0, 1, \ldots, N; \quad i = 0, 1, \ldots, N.$$

Here δ_{ik} is the useful *Kronecker symbol*, taking the value 1 when $i = k$ and the value 0 when $i \neq k$. Although the functions $l_0, l_1, \ldots, l_N$ could be constructed by selecting the vectors $[1, 0, \ldots, 0]^T$, $[0, 1, 0, \ldots, 0]^T$, $\ldots$, $[0, \ldots, 0, 1]^T$ as the right-hand members of Eq. (3.1) in turn, it is geometrically clear what these functions are. They are the *pyramid* or *tent* functions illustrated in Fig. 3.2 and defined by Eq. (3.2) in which j takes the values $1, 2, \ldots, N - 1$:

$$l_0(x) = \begin{cases} \dfrac{x - x_1}{x_0 - x_1}, & a = x_0 \leqslant x \leqslant x_1, \\ 0, & x_1 \leqslant x \leqslant b, \end{cases}$$

$$l_j(x) = \begin{cases} 0, & a \leqslant x \leqslant x_{j-1}. \\ \dfrac{x - x_{j-1}}{x_j - x_{j+1}}, & x_{j-1} \leqslant x \leqslant x_j, \\ \dfrac{x - x_{j+1}}{x_j - x_{j+1}}, & x_j \leqslant x < x_{j+1}, \\ 0, & x_{j+1} \leqslant x \leqslant b, \end{cases} \qquad (3.2)$$

$$l_N(x) = \begin{cases} 0, & a \leqslant x \leqslant x_{N-1}, \\ \dfrac{x - x_{N-1}}{x_N - x_{N-1}}, & x_{N-1} \leqslant x \leqslant x_N = b. \end{cases}$$

In terms of this basis, the interpolant has the simple form

$$l(x) = l_0(x)f_0 + \ldots + l_N(x)f_N$$

$$= [l_0(x), \ldots, l_N(x)] \begin{bmatrix} f_0 \\ \vdots \\ f_N \end{bmatrix}$$

$$= \mathbf{l}(x)^T \mathbf{f},$$

in which the values of the function being interpolated appear explicitly, and it is not necessary to solve equations of the form of Eq. (3.1). Indeed, we may represent *any* function in $\mathscr{L}_N(K)$ by the form

$$l(x) = l_0(x)y_0 + \ldots + l_N(x)y_N = \mathbf{l}(x)^T \mathbf{y}, \qquad (3.3)$$

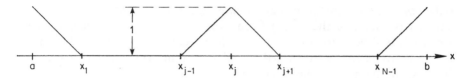

FIG. 3.2   Tent functions.

the values $y_0, \ldots, y_N$ being parameters controlling the form of $l(x)$. This is completely analogous with the Lagrange interpolation process (cf. Section 2.3).

Functions of this kind can also be used to carry out least squares approximation rather than interpolation. In this event one must work with the vector space $\mathscr{L}_M(K)$ in which $1 \leqslant M \leqslant N$, and the only restriction on the knot sequence K is that $k_0 = a$, $k_M = b$, and $k_0 < k_1 < \ldots < k_M$. Thus, the interior knots $k_1, k_2, \ldots, k_{M-1}$ that define the space $l_M(K)$ may bear no relation to the more numerous interior points $x_1, x_2, \ldots, x_{N-1}$ at which data are assigned. This idea will be developed next in Section 3.4.

3.4 Least squares approximation by linear splines

The simplest least squares approximation problem in the present context results when we choose $M = 1$. Then there are no knots between a and b, and the general function $l(x)$ has the form

$$l(x) = \frac{x - x_N}{x_0 - x_N} y_0 + \frac{x - x_0}{x_N - x_0} y_N.$$

This is an arbitrary function of the form $A + Bx$, and so this least squares approximation problem is identical with the usual linear regression problem, as discussed in Chapter 2. Therefore the first case of interest occurs when $M = 2$, and there is an *interior knot* k_1 between a and b. For the time being we assume that the position of the interior knot, or knots, is fixed *a priori*: The approximating function has the form of Eq. (3.3), and the sum of the squares of its deviations from the data f_j at the points x_j, $j = 0, 1, \ldots, N$, is

$$E(l) = \sum_{j=0}^{N} \left\{ \sum_{i=0}^{M} y_i l_i(x_j) - f_j \right\}^2. \tag{3.4}$$

Keep in mind that there are now $N + 1$ distinct points at which data are given and *also* $M + 1$ distinct knots, which may or may not coincide with

data points at which the deviations are calculated. In view of the small
support property of the cardinal functions l_i, many of the values $l_i(x_j)$ in
Eq. (3.4) will be zero. In order for y_i to be actually present in $E(l)$, it is
necessary to locate the knots so that each l_i contains some data in its support
[that is, the points x at which $l_i(x) \neq 0$].

Now the necessary conditions for a minimum of $E(l)$ are

$$\frac{\partial E(l)}{\partial y_k} = 2 \sum_{j=0}^{N} \left\{ \sum_{i=0}^{M} y_i l_i(x_j) - f_j \right\} l_k(x_j) = 0 \quad \text{for} \quad k = 0 \ldots, M. \quad (3.5)$$

The required values of the parameters $y_0, \ldots, y_M$ can be obtained from
these equations, which can be rearranged into the form

$$\begin{bmatrix} a_{00} & \cdots & a_{0M} \\ & \vdots & \\ a_{M0} & \cdots & a_{MM} \end{bmatrix} \begin{bmatrix} y_0 \\ \vdots \\ y_M \end{bmatrix} = \begin{bmatrix} b_0 \\ \vdots \\ b_M \end{bmatrix}, \quad (3.6)$$

where

$$a_{ki} = \sum_{j=0}^{M} l_k(x_j) l_i(x_j), \quad k, i = 0, 1, \ldots, M,$$

$$b_k = \sum_{j=0}^{M} f_j l_k(x_j), \quad k = 0, 1, \ldots, M.$$

In view of the small support property of l_i, the product $l_k l_i = 0$ whenever
$|k - i| \geq 2$, and hence the matrix $[a_{ki}]$ in Eq. (3.6) is *tridiagonal*. This feature
makes the invertibility of the coefficient matrix easy and the solution of
Eq. (3.6) inexpensive.

Concerning the placement of knots and the choice of M, we can follow
the algorithm of Ichida *et al.* (1976) and begin by selecting $M = 2$ and
placing the only interior knot k_1 such that there are nearly as many data
points as possible to the left as to the right of k_1. The best approximation
is computed, and the sum of squared deviations is evaluated to the left and
to the right of k_1. Whichever of the intervals $[a, k]$, $[k, b]$ has the larger
sum is again subdivided by the insertion of a knot. The insertion of knots
is terminated when the value of δ, defined by

$$\delta = E(l)/(N - M), \quad (3.7)$$

where $E(l)$ is as in Eq. (3.4), reaches a plateau, at which point additional
knots do not significantly reduce δ [See Ichida *et al.* (1976) and de Boor
(1978), p. 267]. The quantity δ should be compared with $\hat{\sigma}^2$ defined in Eq.
(2.19).

Figure 3.3 shows a number of approximations to experimentally obtained data, the abscissa being the travel time below the surface of the earth of a sonic signal and the ordinate being the velocity of sound in the rock at that time.

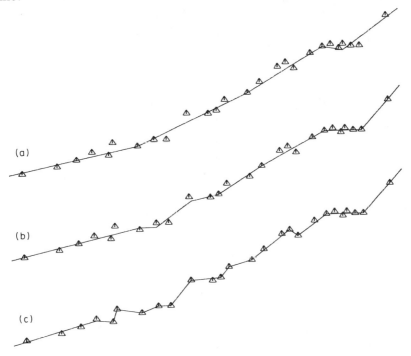

FIG. 3.3   Piecewise linear least squares fits. (a) $M = 5$, $N = 25$, $\delta = 0.005$; (b) $M = 8$, $N = 25$, $\delta = 0.002$; and (c) $M = 15$, $N = 25$, $\delta = 0.001$.

3.5 An alternative construction of the space of linear splines

Clearly, every linear spline l with knot sequence K has a derivative l' that is *piecewise constant*. Typical graphs of a linear spline l together with its derivative l' are shown in Fig. 3.4.

The function l' illustrated in Fig. 3.4 is known as a step function. A very primitive step function is the jump function illustrated in Fig. 3.5. It is defined by

$$J(x) = \begin{cases} -1, & x < 0, \\ 1, & x \geq 0. \end{cases}$$

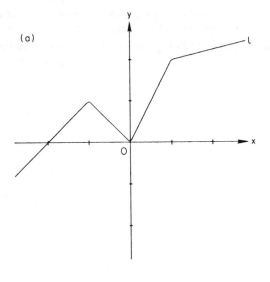

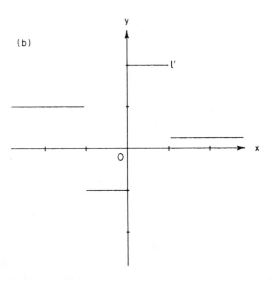

FIG. 3.4   (a) Linear spline l and (b) its derivative l'.

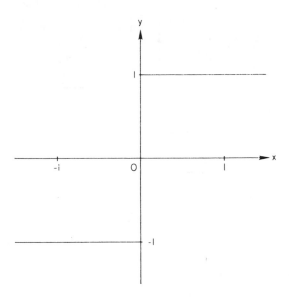

FIG. 3.5   Jump function $J(x)$.

It is important to observe here that $J(x)$ is the derivative of $|x|$, except at $x = 0$, where $|x|$ is not differentiable because of its sharp corner. The jump function can be used to generate more general step functions, such as that in Fig. 3.6. This function has the representation $a + a_i J(x - k_i)$. Observe also that $|x - k_i|$ is an antiderivative or indefinite integral of $J(x - x_i)$. Using this idea, l' can now be expressed in the form

$$l'(x) = A + \sum_{i=1}^{N-1} a_i J(x - k_i),$$

since the first jump occurs at the knot k_1 and the last at k_{N-1}. The values $2|a_i|$, $i = 1, \ldots, N - 1$, are precisely the magnitudes of the jumps. By integration we find that

$$l(x) = Ax + B + \sum_{i=1}^{N-1} a_i |x - k_i|, \tag{3.8}$$

where B is the integration constant. In this expression we see that the number of parameters is now $N + 1$, which confirms the claim of Theorem 3.2.1 that the dimension of $\mathcal{L}_N(K)$ is $N + 1$.

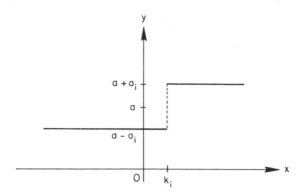

FIG. 3.6   Step function.

3.6 Piecewise cubic functions

Although the nature of the linear spline precludes its being smooth, in the sense of having at least one derivative for every x, it is possible to obtain piecewise *cubic* functions that have at least a continuous first derivative and in some cases, even a continuous second derivative. As in Theorem 3.2.1, let K be the set of knots k_i, satisfying $a = k_0 < k_1 < \ldots < k_N = b$. Let us first observe that, if function values and first derivatives are prescribed at the knots, then a technique for constructing a piecewise cubic interpolating function with continuous first derivative is already at our disposal. The technique developed in Section 2.4 must simply be applied to each segment $[k_{i-1}, k_i]$ for $i = 1, 2, \ldots, N$ [see especially Eqs. (2.8) and (2.9)]. The fact that the same slope is used at k_i for intervals $[k_{i-1}, k_i]$ and $[k_i, k_{i+1}]$ ($i = 1, 2, \ldots, N - 1$) guarantees continuity of slope for the interpolating function at all interior knots.

We shall return to this procedure shortly; initially, a more general approach is to be taken to the construction of piecewise cubics with *at least* one continuous derivative at the interior knots.

Let us construct a representation for any function in the family (in fact, the vector space) of all functions that are ordinary cubic polynomials between consecutive knots and are not only continuous but also possess a continuous derivative on $[a, b]$, i.e. they are to belong to $\mathscr{C}^1[a, b]$. If $S(x)$ is such a function, then its *second* derivative has the form

$$S''(x) = l(x) + \sum_{i=1}^{N-1} b_i J(x - k_i), \tag{3.9}$$

in which $l(x)$ is as in Section 3.5. The justification for this form for $S''(x)$ is that $S(x)$ is assumed to be piecewise cubic and at least once differentiable at the knots. Hence its second derivative is piecewise *linear*, though not necessarily continuous at the knots. Now $l(x)$ is piecewise linear and continuous, and the term with the jump functions serves to introduce jumps of as yet arbitrary size $2|b_i|$ at the knots.

By integrating Eq. (3.9) twice, taking advantage of the expression (3.8) for $l'(x)$ and denoting the integration constants by C and D, we obtain

$$S(x) = \frac{Ax^3}{3} + \frac{Bx^2}{2} + Cx + D + \sum_{i=1}^{N-1} a_i \tfrac{1}{6}|x - k_i|^3$$

$$+ \sum_{i=1}^{N-1} b_i \tfrac{1}{2}|x - k_i|^2 J(x - k_i). \tag{3.10}$$

Here we have also used the facts that $|x|$ is the second derivative of $\tfrac{1}{6}|x|^3$, and $J(x)$ is the second derivative of $\tfrac{1}{2}|x|^2 J(x)$. It follows that a typical function $\tfrac{1}{6}|x - k_i|^3$ under the first summation is twice continuously differentiable at k_i; its second derivative is simply $|x - k_i|$, which is continuous at k_i. On the other hand, the function $\tfrac{1}{2}|x - k_i|^2 J(x - k_i)$ has only a *first* derivative at k_i. We see that $S(x)$ contains $2(N - 1) + 4 = 2N + 2$ parameters.

Theorem 3.6.1 *Let K denote the set of knots $\{k_0, \ldots, k_N\}$, with $a = k_0 < k_1 < \ldots < k_N = b$. The set of all (piecewise cubic) functions of the form of Eq. (3.10) is a vector space of dimensions $2N + 2$.*

We denote the space of Theorem 3.6.1 by $\mathscr{C}_N(K)$ and note that it is a subspace of $\mathscr{C}^1[a, b]$.

The fact that the space $\mathscr{C}_N(K)$ has dimension $2N + 2$ can be made plausible by the following argument: The knot sequence K defines N subintervals and a cubic polynomial having four coefficients is defined on each of these, so there are $4N$ coefficients in all. To obtain $\mathscr{C}^1$ continuity, we apply two constraints at each *interior* knot—a total of $2N - 2$ constraints. The number of coefficients minus the number of constraints is simply $2N + 2$, the dimension of $\mathscr{C}_N(K)$.

3.7 Interpolation with piecewise cubic functions

Clearly, the dimension of $\mathscr{C}_N(K)$ is too large to permit unique interpolation if, as is usual, the abscissas x_i of the data coincide with the knots k_i. If, however, the slopes m_i of the interpolant are also specified at the points x_i, then the number of conditions of interpolation is equal to the number of free parameters in a typical function $S \in \mathscr{C}_N(K)$ and, furthermore, a unique interpolant exists.

Theorem 3.7.1   *There exists a unique function S in $\mathscr{C}_N(K)$ that has prescribed values f_i and prescribed slopes m_i, $i = 0, 1, \ldots, N$, at the knots k_i.*

We shall examine the nature of this interpolation scheme in a moment, but first let us make a further comment on Eq. (3.10) that will be important in Chapter 4. Theorem 3.7.1 leads to the construction of an interpolant in $\mathscr{C}[a, b]$, and, although the form [Eq. (3.10)] of $S(x)$ is not convenient for computation, it shows explicitly the conditions under which $S(x)$ will be *twice* differentiable. For, if there are to be no discontinuities in S'', then all the coefficients $b_i = 0$ in Eqs. (3.6.1) and (3.6.2). In this case, the resulting $S(x)$ is called a *cubic spline*. (Unless otherwise stated, it is understood throughout this book that a cubic spline is twice differentiable at the knots.) Denote by $\mathscr{S}_N(K)$ the family of such functions. Thus $S(x)$ is a cubic spline on the set of knots K if and only if it has the form

$$S(x) = \tfrac{1}{6}Ax^3 + \tfrac{1}{2}Bx^2 + Cx + D + \tfrac{1}{6}\sum_{i=1}^{N-1} a_i|x - k_i|^3, \qquad (3.11)$$

for some constants $A, B, C, D, a_1, a_2, \ldots, a_{N-1}$.

Theorem 3.7.2   *If K is a set of $N + 1$ distinct knots on $[a, b]$, then the family $\mathscr{S}_N(K)$ is a vector space of dimension $N + 3$ and is a subspace of $\mathscr{C}^2[a, b]$.*

We conclude that *unique* interpolation at $N + 1$ knots in $[a, b]$ with cubic splines is also not possible. There remain two degrees of freedom. We shall pursue this topic in Chapter 4 and concentrate in the remainder of this chapter on piecewise cubics with just one continuous derivative. As in the case of piecewise linear interpolation, the use of cardinal functions leads to very convenient and transparent representation of interpolants.

Consider interpolation in the context of Theorem 3.7.1, in which the knots $k_i = x_i$, and two distinct types of data are supplied at each knot. Thus it is necessary to construct $2N + 2$ cardinal functions. Theorem 3.7.1 assures us of the existence of unique functions Φ_i, Ψ_i, $i = 0, 1, \ldots, N$, in $\mathscr{C}_N(K)$ with the properties

$$\Phi_i(x_j) = \delta_{ij}, \quad \Phi_i'(x_j) = 0, \quad \Psi_i(x_j) = 0, \quad \Psi_i'(x_j) = \delta_{ij}. \quad (3.12)$$

for $i, j = 0, 1, \ldots, N$. Each of these functions has prescribed values and slopes at the knots. Now an interpolant can be written immediately:

$$S(x) = \sum_{i=0}^{N} \Phi_i(x)f_i + \sum_{i=0}^{N} \Psi_i(x)m_i. \qquad (3.13)$$

The precise form of these cardinal functions could be obtained from Eq. (3.10) by imposing the conditions in Eq. (3.12). However, it is easier to exploit the fact that the Φ_i and Ψ_i are piecewise cubic polynomials and can be constructed with the aid of the technology of Section 2.4.

It has been shown in Section 2.4 that a cubic is uniquely determined by its values and slopes at the ends of an interval. Consequently, for $i = 1, 2, \ldots, N - 1$, $\Phi_i(x) \equiv 0$ when $x < x_{i-1}$ or $x > x_{i+1}$ (see Fig. 3.7).

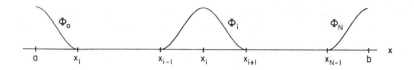

FIG. 3.7   Cardinal functions Φ_i.

Also, $\Phi_0(x) = 0$ when $x > x_1$, and $\Phi_N(x) \equiv 0$ when $x < x_{N-1}$. Hence the Φ_i enjoy the small support property. Now, for $i = 1, \ldots, N - 1$, Φ_i on $[x_{i-1}, x_{i+1}]$ is composed of left and right segments; these are essentially the H_0 and H_1 functions of Section 2.4. Several of these cardinal functions are shown in Fig. 3.7. Thus, for $i = 1, \ldots, N - 1$, with $h_i = x_i - x_{i-1}$,

$$\Phi_i(x) = \begin{cases} 0, & x < x_{i-1}, \\ (2/h_i^3)(x - x_{i-1})^2(x - x_i - \tfrac{1}{2}h_i), & x_{i-1} \leqslant x < x_i, \\ (2/h_{i+1}^3)(x - x_i + \tfrac{1}{2}h_{i+1})(x - x_{i+1})^2, & x_i \leqslant x < x_{i+1}, \\ 0, & x \geqslant x_{i+1}. \end{cases} \quad (3.14)$$

For $i = 0$, only the last two parts of this definition of Φ_i apply, whereas if $i = N$, only the first two parts are needed.

FIG. 3.8   Cardinal functions Ψ_i.

In a similar way we can determine the Ψ_i, shown in Fig. 3.8, from K_0 and K_1 of Section 2.4. For $i = 1, \ldots, N - 1$,

$$\Psi_i(x) = \begin{cases} 0, & x < x_{i-1}, \\ (h_i^2)^{-1}(x - x_{i-1})^2(x - x_i), & x_{i-1} \leqslant x < x_i, \\ (h_{i+1}^2)^{-1}(x - x_i)(x - x_{i+1})^2, & x_i \leqslant x < x_{i+1}, \\ 0, & x \geqslant x_{i+1}. \end{cases} \quad (3.15)$$

When $i = 0$ or N, the remarks concerning Φ_0 and Φ_N apply.

We emphasize here that, although they appear visually quite smooth, these cardinal functions are only in $\mathscr{C}^1[x_0, x_N]$ and not in $\mathscr{C}^2[x_0, x_N]$. In many applications, only the values f_i are available. The slopes m_i are then parameters whose values can be manipulated in order to adjust the shape of the interpolating curve.

In some interpolation methods, the slopes are determined from the ordinates in a *linear* way. That is, the method involves a matrix M such that

$$
\begin{bmatrix} m_0 \\ \vdots \\ m_N \end{bmatrix} = M \begin{bmatrix} f_0 \\ \vdots \\ f_N \end{bmatrix}
\tag{3.16}
$$

Then we can write $S(x)$ [cf. Eq. (3.13)] in the form,

$$
S(x) = [\Phi_0(x), \ldots, \Phi_N(x)] \begin{bmatrix} f_0 \\ \vdots \\ f_N \end{bmatrix} + [\Psi_0(x), \ldots, \Psi_N(x)] \begin{bmatrix} m_0 \\ \vdots \\ m_N \end{bmatrix}
$$

$$
= ([\Phi_0(x), \ldots, \Phi_N(x)] + [\Psi_0(x), \ldots, \Psi_N(x)]M) \begin{bmatrix} f_0 \\ \vdots \\ f_N \end{bmatrix}
\tag{3.17}
$$

$$
= [\phi_0(x), \ldots, \phi_N(x)] \begin{bmatrix} f_0 \\ \vdots \\ f_N \end{bmatrix}
$$

where we have put

$$
\phi_i(x) = \Phi_i(x) + [\Psi_0(x), \ldots, \Psi_N(x)]M_i,
\tag{3.18}
$$

and M_i is the ith column of M.

The matrix M thus makes the interpolant unique and determines a corresponding set of cardinal functions ϕ_i, $i = 0, \ldots, N$. The function $\phi_i(x)$ will not have the small support feature unless M_i has non-zero elements only near the ith row. Even if the small support is lost, it is important that $\phi_i(x)$ at least attenuate strongly as x moves away from x_i. The reason for this is revealed by examining the effect on $S(x)$ of a perturbation ϵ in the value of one ordinate, say f_k. Then the perturbed interpolant is

$$
S_\epsilon(x) = \sum_{\substack{i=0 \\ i \neq k}}^{N} f_i \phi_i(x) + (f_k + \epsilon)\phi_k(x) = S(x) + \epsilon\phi_k(x).
$$

Consequently, the effect of the perturbation at x_k is localized near x_k only when ϕ_k attenuates rapidly.

An especially important choice of M is that which renders $S(x)$ twice differentiable. There is a two-parameter family of such matrices in view of the dimensionality comment in Theorem 3.7.2. It then turns out that M^{-1} can be tridiagonal, but M itself is "full", with the result that the cardinal functions of the resulting *cubic spline* do not have small support. They do, however, have good attenuation. Cubic splines are the subject of Chapter 4.

It is also possible to make non-linear choices for the determination of $(m_0, \ldots, m_N)$ in the form

$$\begin{bmatrix} m_0 \\ \vdots \\ m_N \end{bmatrix} = F(f_0, \ldots, f_N), \tag{3.19}$$

where F is a vector-valued function of $f_0, \ldots, f_N$. In that event the cardinal functions $\phi_i(x)$ can no longer be used for a simple representation of an interpolant. We present some linear and non-linear schemes next in Section 3.8.

3.8 Choice of slopes in piecewise cubic interpolation

In the case of computer aided design, numerical values can be assigned to the slopes, the interpolating function can be examined and then modified in shape by readjusting the slopes. Clearly, a reasonable initial choice is necessary and can be made on the basis of the schemes discussed below. The result may then be accepted or rejected on the basis of visual and frequently unquantifiable criteria.

If a linear scheme of the form of Eq. (3.16) is desired, then perhaps the simplest choice for m_i, $i = 1, \ldots, N - 1$, is the slope at x_i of the unique parabola that interpolates at x_{i-1}, x_i, x_{i+1}. This is easily found to be

$$m_i = \frac{[(f_{i+1} - f_i)/h_{i+1}]h_i + [(f_i - f_{i-1})/h_i]h_{i+1}}{h_{i+1} + h_i},$$

where $h_i = x_i - x_{i-1}$. For end slopes we may use

$$m_0 = (f_1 - f_0)/h_1, \qquad m_N = (f_N - f_{N-1})/h_N.$$

It is convenient to set

$$\lambda_i = h_{i+1}/(h_i + h_{i+1}), \quad \mu_i = 1 - \lambda_i = h_i/(h_i + h_{i+1}), \qquad i = 1, \dots, N-1,$$

and $\lambda_0 = 0$, $\lambda_N = 1$, $\mu_0 = 1$, $\mu_N = 0$. Then

$$m_i = [(f_{i+1} - f_i)/h_{i+1}]\mu_i + [(f_i - f_{i-1})/h_i]\lambda_i, \qquad i = 0, 1, \dots, N.$$

The associated matrix M is

$$
\begin{bmatrix}
-\dfrac{1}{h_1} & \dfrac{1}{h_1} & 0 & 0 & . & . & 0 & 0 & 0 \\[2mm]
-\dfrac{1}{h_1} & \left[\dfrac{\lambda_1}{h_1} - \dfrac{\mu_1}{h_2}\right] & \dfrac{\mu_1}{h_2} & 0 & . & . & . & 0 & 0 \\[2mm]
0 & -\dfrac{\lambda_2}{h_2} & \left[\dfrac{\lambda_2}{h_2} - \dfrac{\mu_2}{h_3}\right] & \dfrac{\mu_2}{h_3} & . & & & . & 0 \\[2mm]
0 & 0 & & . & . & . & & . & . \\
. & . & & . & . & . & . & . & . \\
0 & . & & . & . & 0 & -\dfrac{\lambda_{N-2}}{h_{N-2}} & \left[\dfrac{\lambda_{N-2}}{h_{N-2}} - \dfrac{\mu_{N-2}}{h_{N-1}}\right] & \dfrac{\mu_{N-2}}{h_{N-1}} & 0 \\[2mm]
0 & 0 & . & . & . & 0 & -\dfrac{\lambda_{N-1}}{h_{N-1}} & \left[\dfrac{\lambda_{N-1}}{h_{N-1}} - \dfrac{\mu_{N-1}}{h_N}\right] & \dfrac{\mu_{N-1}}{h_N} \\[2mm]
0 & 0 & 0 & . & . & . & 0 & -\dfrac{1}{h_N} & \dfrac{1}{h_N}
\end{bmatrix}
$$

Since M is tridiagonal, the cardinal function $\phi_i(x)$ has the form

$$\phi_i(x) = \Phi_i(x) + m_{i-1,i}\Psi_{i-1}(x) + m_{i,i}\Psi_i(x) + m_{i+1,i}\Psi_{i+1}(x),$$

$$i = 0, 1, \dots, N,$$

where $m_{i,j}$ is the element in row i, column j of M, and we use the convention that $m_{i,j}$ with some subscripts outside the range $1, \dots, N$ are set equal to zero. Thus, the support of $\phi_i(x)$ is $[x_{i-2}, x_{i+2}]$ when $i = 2, \dots, N - 2$ and somewhat smaller when $i = 0, 1, N - 1, N$, making this a *local method*. It is sometimes referred to as Bessel's method.

A non-linear choice, suggested by Akima (1970), is

$$m_i = \frac{|S_{i+2} - S_{i+1}|S_i + |S_i - S_{i-1}|S_{i+1}}{|S_{i+2} - S_{i+1}| + |S_i - S_{i-1}|},$$

where S_k is the slope of the line segment joining the points (x_{k-1}, f_{k-1}) and (x_k, f_k) for $k = i - 1, i, i + 1, i + 2$.

When $i = 0$ or N, m_i requires the provision of an auxiliary point outside the interval. It is also possible to choose M so that the curve $S(x)$ has a continuous second derivative. The resulting function is the classical cubic spline which is to be discussed in Chapter 4.

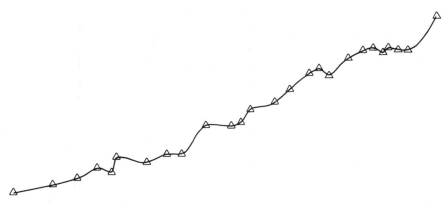

Fig. 3.9 Akima's method.

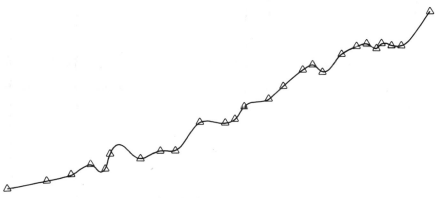

Fig. 3.10 Bessel's method.

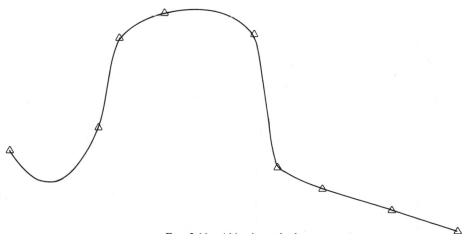

Fig. 3.11 Akima's method.

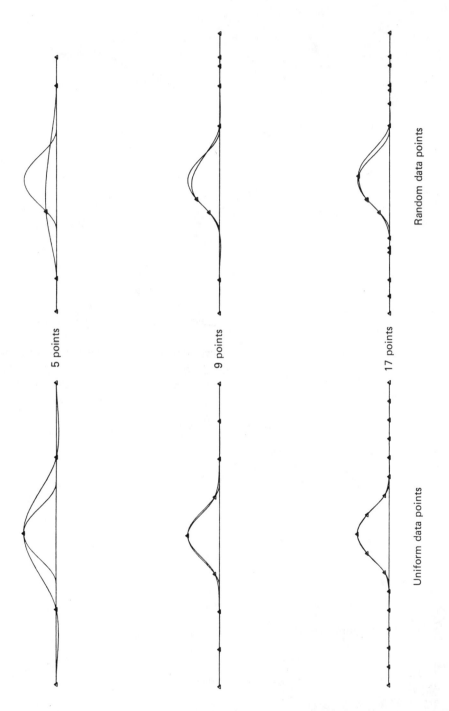

Uniform data points   Random data points

5 points

9 points

17 points

FIG. 3.12 Piecewise interpolants: mountain.

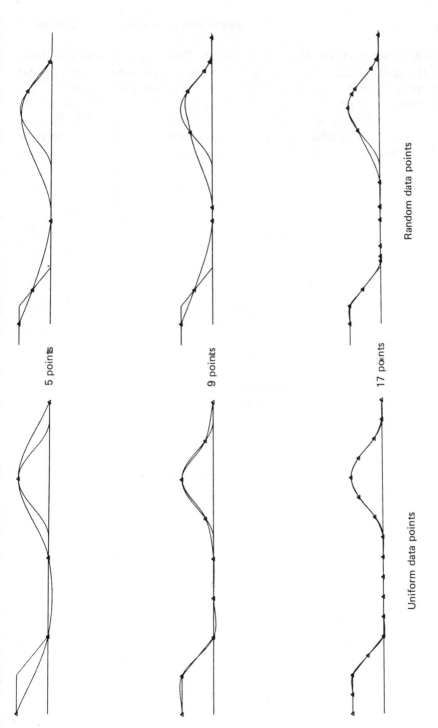

5 points

9 points

17 points

Uniform data points

Random data points

Fig. 3.13   Piecewise interpolants: ramp plus mountain.

In Figs. 3.9–3.11, we show interpolants by Bessels' and Akima's methods for the data of Fig. 3.3 as well as for other data, which exhibit great variation and can cause difficulties for many interpolation methods.

Figures 3.12 and 3.13 show the piecewise cubic interpolants to the mountain and ramp–mountain data introduced in Section 2.5. Bessel's method is used to obtain the necessary slopes, and both the underlying function and interpolant are shown.

4

Curve Fitting with Cubic Splines

4.1 Introduction

It has been indicated in Chapter 3 that in many curve-fitting problems, curves constructed as "piecewise" polynomials have significant advantages over simple polynomials of high degree. There are, of course, many ways of forming piecewise polynomial fits, but the history of the past several years shows that *cubic splines* take pride of place as the most widely useful. Recall that, with a given knot sequence K, $a = k_0 < k_1 < \ldots < k_N = b$, a cubic spline $S(x)$ is a cubic polynomial in each subinterval $[k_0, k_1]$, $[k_1, k_2]$, $\ldots, [k_{N-1}, k_N]$, but these cubic segments of the function are joined together at the interior knots $k_1, k_2, \ldots, k_{N-1}$ in such a way that $S(x)$ has *two* continuous derivatives on $[a, b]$, i.e. so that $S(x) \in \mathscr{C}^2[a, b]$.

It is easily verified that if $x_0 \in (a, b)$, the function

$$f(x) = |x - x_0|^3$$

is in $\mathscr{C}^2[a, b]$. Indeed, we can compute

$$f'(x) = \begin{cases} 3(x - x_0)^2 & \text{for } x \geq x_0, \\ -3(x - x_0)^2 & \text{for } x \leq x_0, \end{cases} \qquad f''(x) = 6|x - x_0|,$$

and observe that $f''(x)$ is continuous on $[a, b]$. In particular, $f''(x_0)$ is well defined and has the value zero. In Fig. 4.1 we sketch f, f', and f''.

It has been shown in Section 3.7 that every cubic spline can be written in the form

$$S(x) = \alpha x^3 + \beta x^2 + \gamma x + \delta + \sum_{i=1}^{N-1} a_i |x - k_i|^3, \qquad (4.1)$$

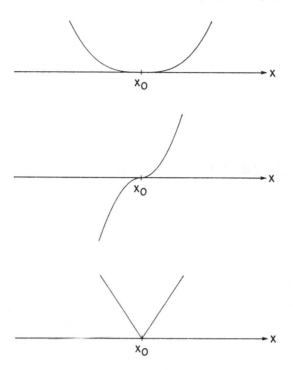

FIG. 4.1   Function $f(x) = |x - x_0|^3$ and its derivatives.

[cf. Eq. (3.11)] for some constants α, β, γ, δ, a_1, a_2, ..., a_{N-1}. With the above observation, it is clear that $S(x) \in \mathscr{C}^2[a, b]$ and also that $S(x)$ is a cubic polynomial on each subinterval $[k_i, k_{i+1}]$ for $i = 0, 1, \ldots, N - 1$. Indeed, $S(x)$ is simply a linear combination of the functions

$$x^3, \quad x^2, \quad x, \quad 1, \quad |x - k_1|^3, \quad \ldots, \quad |x - k_{N-1}|^3,$$

i.e. any $S(x)$ can be formed by multiplying each of these functions by a constant and summing, as in Eq. (4.1).

In Chapter 2 we have emphasized the point that, although a polynomial function in $\mathscr{P}_N$ is a linear combination of the primitive functions 1, x, x^2, ..., x^N, in finding a particular polynomial $p(x) = \Sigma_{i=0}^{N} a_i x^i$, which fits certain data, it is generally *not* a good idea to try to calculate the coefficients $a_0, a_1, \ldots, a_N$ directly. The same applies to the representation of Eq. (4.1) for a cubic spline.* Thus, the Sections 4.2 and 4.3 are devoted to the presentation of two very different approaches to the representation and computation of cubic splines.

* For the same reason we have also omitted a discussion of the so-called truncated power functions for the representation of a cubic spline.

4.2. Finding slopes from function values

Our first approach to the representation of interpolating cubic splines takes advantage of Eq. (3.13) in which any piecewise cubic function defined on $[a, b]$, which is also in $\mathscr{C}^1[a, b]$ (note only *one* continuous derivative), is written in terms of the $2N + 2$ primitive cardinal functions

$$\Phi_0, \Phi_1, \ldots, \Phi_N, \Psi_0, \Psi_1, \ldots, \Psi_N$$

defined by Eq. (3.12). The representation (3.13) contains $2N + 2$ parameters and the condition that $S(x)$ takes the values $f_0, f_1, \ldots, f_N$ at k_0, $k_1, \ldots, k_N$, respectively, applies $N + 1$ constraints. The remaining $N + 1$ constraints include the $N - 1$ conditions that $S''(x)$ be continuous at the interior knots $k_1, k_2, \ldots, k_{N-1}$. These ensure that $S(x) \in \mathscr{C}^2[a, b]$ and not merely $\mathscr{C}^1[a, b]$.

We show first that these higher smoothness constraints allow us to express the slope parameters m_i in Eq. (3.13) *explicitly* in terms of the function values $f_0, f_1 \ldots, f_N$. Once this is done, the spline is determined for all $x \in [a, b]$ by means of Eq. (3.13). As in this equation, we assume that $k_i = x_i, i = 0, 1, \ldots, N$.

For a general function with the form of Eq. (3.13), we first calculate the jumps in second derivative at the interior knots $k_1, \ldots, k_{N-1}$ and equate these to zero. We do this by calculating the difference in values between $S''(x_k)$ computed from the left segment of the piecewise cubic and denoted by $S''(x_k^-)$,[*] and the similar quantity $S''(x_k^+)$ obtained from the segment to the right of x_k. Thus,

$$S''(x_k^-) - S''(x_k^+) \tag{4.2}$$

$$= \sum_{i=0}^{N} f_i[\Phi_i''(x_k^-) - \Phi_i''(x_k^+)] + \sum_{i=0}^{N} m_i[\Psi_i''(x_k^-) - \Psi_i''(x_k^+)]$$

$$= \sum_{i=k-1}^{k+1} f_i[\Phi_i''(x_k^-) - \Phi_i''(x_k^+)] + \sum_{i=k-1}^{k+1} m_i[\Psi_i''(x_k^-) - \Psi_i''(x_k^+)] = 0.$$

The sums have three terms each because of the small support of the cardinal functions. When the indicated computations are carried out, we find that the $N + 1$ slopes satisfy the $N - 1$ equations

$$\frac{1}{h_k} m_{k-1} + 2\left(\frac{1}{h_k} + \frac{1}{h_{k+1}}\right) m_k + \frac{1}{h_{k+1}} m_{k+1} = 3\frac{f_k - f_{k-1}}{h_k^2} + 3\frac{f_{k+1} - f_k}{h_{k+1}^2},$$

$$k = 1, \ldots, N - 1. \tag{4.3}$$

[*] In the notation of Section 1.2, $S''(x_k^-) = \lim_{x \to x_k^-} S''(x)$ and $S''(x_k^+) = \lim_{x \to x_k^+} S''(x)$.

There remain two degrees of freedom, as claimed in Theorem 3.7.2. It is customary to select values for m_0 and m_N in order to define the remaining slopes uniquely, or more generally, one may supply two more equations involving m_0 and m_N. These are normally written in the form

$$2m_0 + \mu_0 m_1 = c_0, \qquad \lambda_N m_{N-1} + 2m_N = c_N, \qquad (4.4)$$

where μ_0, λ_N, c_0, and c_N are parameters at our disposal.

With the notation

$$\lambda_k = h_{k+1}/(h_k + h_{k+1}), \qquad\qquad\qquad \mu_k = 1 - \lambda_k,$$
$$c_k = 3\lambda_k[(f_k - f_{k-1})/h_k] + 3\mu_k[(f_{k+1} - f_k)/h_{k+1}], \quad k = 1, \ldots, N-1, \qquad (4.5)$$

Eqs. (4.3) and (4.4) can be written in the form

$$
\begin{bmatrix}
2 & \mu_0 & 0 & \cdots & 0 & 0 & 0 \\
\lambda_1 & 2 & \mu_1 & \cdots & 0 & 0 & 0 \\
0 & \lambda_2 & 2 & \cdots & 0 & 0 & 0 \\
 & & & & & & \\
0 & 0 & 0 & \cdots & 2 & \mu_{N-2} & 0 \\
0 & 0 & 0 & \cdots & \lambda_{N-1} & 2 & \mu_{N-1} \\
0 & 0 & 0 & \cdots & 0 & \lambda_N & 2
\end{bmatrix}
\begin{bmatrix}
m_0 \\ m_1 \\ m_2 \\ \vdots \\ m_{N-2} \\ m_{N-1} \\ m_N
\end{bmatrix}
=
\begin{bmatrix}
c_0 \\ c_1 \\ c_2 \\ \vdots \\ c_{N-2} \\ c_{N-1} \\ c_N
\end{bmatrix}. \qquad (4.6)
$$

More details concerning the derivation of this system can be found in Ahlberg et al. (1967), who also give a convenient algorithm for solving this tridiagonal system. This algorithm can also be used for the least squares : linear spline.

Concerning Eqs. (4.4), various end conditions can be imposed on the spline by specializing the values of μ_0, λ_N, c_0, and c_N. For example, if the numerical value of the slope of $S(x)$ at the left end is given, then we set $\mu_0 = 0$ and $c_0 = 2S'(x_0)$ to obtain $m_0 = S'(x_0)$. A corresponding strategy at the right end, when $S'(x_N)$ is known, is to set $\lambda_N = 0$ and $c_N = 2S'(x_N)$.

The two extra freedoms of the interpolating cubic spline can also be used to specify the *second* derivative of $S(x)$ at the two end points. If the second derivative of the spline is given at x_0, the choices $\mu_0 = \lambda_N = 1$, and appropriate choices for c_0 and c_N yield, at the left end,

$$2m_0 + m_1 = 3[(f_1 - f_0)/h_1] - (h_1/2)S''(x_0), \qquad (4.7)$$

and at the right end,

$$m_{N-1} + 2m_N = 3[(f_N - f_{N-1})/h_N] + (h_N/2)S''(x_N). \qquad (4.8)$$

A particularly important spline is obtained on setting the second derivatives (and hence the curvatures) equal to zero at the end points.

Definition *A natural cubic spline S is a cubic spline with the additional properties* $S''(x_0) = S''(x_N) = 0.$

Accordingly, an interpolating natural cubic spline satisfies Eq. (4.6) in which the first and last equations are simply Eqs. (4.7) and (4.8), with $S''(x_0) = S''(x_N) = 0$. In other words, it satisfies Eq. (4.6) together with

$$\mu_0 = 1, \quad c_0 = 3[(f_1 - f_0)/h_1]; \quad \lambda_N = 1, \quad c_N = 3[(f_N - f_{N-1})/h_N].$$

In Theorem 4.2.1, we summarize some of the most common end conditions giving rise to a unique interpolating cubic spline. Note that the natural cubic spline interpolant is covered by case (2) (on putting $S''(x_0) = S''(x_N) = 0$). It is often necessary to construct interpolants for periodic data, in which case the interpolant should also be periodic. This can be achieved in the following way: The end conditions and ordinates are determined in such a way that if $S(x)$ is translated through the distance $x_N - x_0$ to the left or right, the translates join up in a $\mathscr{C}^2$ way at x_0 and x_N. This means that $S(x)$ satisfies the *periodic end conditions*

$$f_0 = f_N = S(x_0) = S(x_N), \quad S'(x_0) = S'(x_N), \quad S''(x_0) = S''(x_N),$$

together with the periodicity condition

$$S(x + n(x_N - x_0)) = S(x), \quad n = 0, \pm 1, \pm 2, \ldots .$$

Theorem 4.2.1 *There exists a unique cubic spline satisfying the conditions of interpolation $S(x_j) = f_j$ for $j = 0, 1, \ldots, N$ together with one of the following sets of end conditions*:

(1) *prescribed slopes m_0, m_N at x_0, x_N, respectively*;
(2) *prescribed second derivatives $S''(x_0)$ and $S''(x_N)$*;
(3) *end conditions of the form of Eq. (4.4) with $\lambda_N < 4$ and $\mu_0 < 4$*;
(4) *periodic end conditions.*

In Fig. 4.2 we illustrate a cardinal natural cubic spline. The rapid attenuation is evident. Natural cubic spline interpolants to the mountain and ramp–mountain data are shown in Figs. 4.3 and 4.4, and should be compared with the polynomial interpolants in Figs. 2.3 and 2.4.

Rather than impose end conditions exclusively, one or the other can be replaced by a *not-a-knot* condition at x_1 or x_{N-1} [see de Boor (1978)]. This is simply the requirement that the *third* derivative be continuous at the selected knot. The result of this is that the two cubic segments that join at the knot are portions of one and the same cubic polynomial, which is $\mathscr{C}^\infty$,

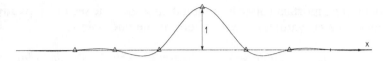

FIG. 4.2   Cardinal natural cubic spline.

and the knot has effectively vanished. The appropriate equations relating the slopes are obtained by evaluating Eq. (4.2) with third, instead of second, derivatives, and then using the second and next-to-last equation from Eq. (4.6) to eliminate m_2 and m_{N-2}, respectively. The resulting choice for the parameters μ_0, c_0, λ_N, c_N in Eq. (4.4) is then

$$\mu_0 = \frac{2(h_1 + h_2)}{h_2}, \qquad c_0 = \frac{2(3h_1 + 2h_2)(f_1 - f_0)}{(h_1 + h_2)h_1} + \frac{2h_1^2(f_2 - f_1)}{h_2(h_1 + h_2)},$$

for third-derivative continuity at x_1, and

$$\lambda_N = \frac{2(h_{N-1} + h_N)}{h_{N-1}},$$

$$c_N = \frac{2h_N^2(f_{N-1} - f_{N-2})}{h_{N-1}(h_{N-1} + h_N)} + \frac{2(2h_{N-1} + 3h_N)(f_N - f_{N-1})}{(h_{N-1} + h_N)h_N},$$

for third-derivative continuity at x_{N-1}.

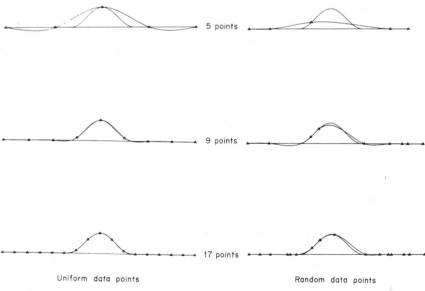

5 points

9 points

17 points

Uniform data points   Random data points

FIG. 4.3   Natural cubic spline interpolants: mountain.

Using the definitions of Eq. (4.5), the right-hand side of the system (4.6) can be written in the matrix form,

$$
\begin{bmatrix} c_0 \\ c_1 \\ c_2 \\ \vdots \\ c_{N-2} \\ c_{N-1} \\ c_N \end{bmatrix} = 3 \begin{bmatrix} 0 & 0 & 0 & 0 & \cdots & 0 \\ -\dfrac{\lambda_1}{h_1} & \dfrac{\lambda_1}{h_1} - \dfrac{\mu_1}{h_2} & \dfrac{\mu_1}{h_2} & 0 & \cdots & 0 \\ 0 & & & & & \\ \vdots & & & & & \\ & & -\dfrac{\lambda_{N-1}}{h_{N-1}} & \dfrac{\lambda_{N-1}}{h_{N\ 1}} - \dfrac{\mu_{N+1}}{h_N} & \dfrac{\mu_{N+1}}{h_N} & \\ 0 & \cdots & & 0 & & 0 \end{bmatrix} \begin{bmatrix} f_0 \\ f_1 \\ f_3 \\ \vdots \\ f_{N-2} \\ f_{N-1} \\ f_N \end{bmatrix} + \begin{bmatrix} \alpha \\ 0 \\ 0 \\ \vdots \\ 0 \\ 0 \\ \beta \end{bmatrix}, \quad (4.9)
$$

α and β being determined by the specific end conditions of the form (4.4) that are selected. If the slopes $S'(x_0)$ and $S'(x_N)$ of the interpolating spline are created linearly from the function values, e.g. if $S'(x_0) = (f_1 - f_0)/h_1$ and $S'(x_N) = (f_N - f_{N-1})/h_N$, then the vector $[\alpha, 0, \ldots, 0, \beta]^T$ above can be absorbed into the matrix-vector product by changing the first row of the matrix to $[-\frac{2}{3}(1/h_1), \frac{2}{3}(1/h_1), 0, \ldots, 0]$ and the last row to $[0, \ldots, 0, -\frac{2}{3}(1/h_N), \frac{2}{3}(1/h_N)]$. If a natural cubic spline is desired then in

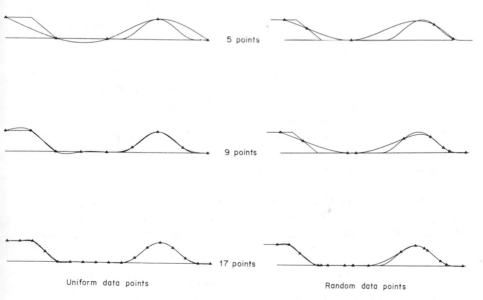

Uniform data points   Random data points

FIG. 4.4   Natural cubic spline interpolants: ramp plus mountain.

view of Eqs. (4.7) and (4.8), the top row becomes $[-1/h_1, 1/h_1, 0, \ldots, 0]$, and the last row is $[0, \ldots, 0, -1/h_N, 1/h_N]$. In both of these cases, Eq. (4.9) takes the form

$$\mathbf{c} = C\mathbf{f},$$

with an appropriate choice of the square matrix C. Then we may write

$$\mathbf{m} = A^{-1}C\mathbf{f},$$

where A denotes the coefficient matrix in Eq. (4.6). Thus, the matrix M of Eq. (3.15) is given by

$$M = A^{-1}C. \qquad (4.10)$$

Even though A is tridiagonal, this product is full, and hence the associated cardinal splines, which could be constructed by setting $\mathbf{f} = (1, 0, \ldots, 0)$, $(0, 1, 0, \ldots, 0), \ldots, (0, 0, \ldots, 0, 1)$ in turn in Eq. (3.17), are not of small support. Because the small-support property is advantageous for computation, one is led to enquire whether there exist non-cardinal spline basis functions having minimal support. That is indeed the case—the basis functions are known as B-splines.

4.3 *B*-splines

We have seen that the tent functions of Section 3.3 form a convenient basis for $\mathcal{L}_N(K)$, the vector space of linear splines. We are now to construct an analogous basis for the space $\mathcal{S}_N(K)$ of cubic splines. The underlying idea is the generation of a basis of minimal support.

We observe that the support of the tent functions that belong to interior knots is two intervals. As well, the functions l_0 and l_N can be viewed as restrictions to $[x_0, x_N]$ of tent functions with support $[x_{-1}, x_1]$ and $[x_{N-1}, x_{N+1}]$, respectively, x_{-1} and x_{N+1} being exterior knots that play no role in the interpolation process.

Consider now the parallel situation with cubic splines. It turns out that the (twice differentiable) smoothness of a cubic spline requires it to have a support of at least four consecutive intervals.

Theorem 4.3.1 *Let K be a knot sequence satisfying $k_0 < k_1 < \ldots < k_M$. For $j = 2, 3, \ldots, M - 2$, there exists a choice of non-zero ordinates f_{j-1}, f_j, f_{j+1} such that the natural cubic spline with knot sequence K that satisfies*

$$S(k_i) = \begin{cases} f_i, & i = j - 1, j, j + 1, \\ 0, & \text{otherwise,} \end{cases}$$

vanishes outside the interval (k_{j-2}, k_{j+2}) (and has zero slope at k_{j-2}, k_{j+2}).

We denote such a B-spline by B_{j-2}. A spline with the property of Theorem 4.3.1 is shown in Fig. 4.5.

We shall demonstrate this theorem in the case when the knots are equally spaced h units apart and when $M = 4$. In Eq. (4.6), set $N = M = 4$, $m_0 = m_4 = 0$, $f_0 = f_4 = 0$, and select the natural spline end conditions via Eqs. (4.7) and (4.8), and Definition 4.2.1. The right-hand side of Eq. (4.6) can be calculated by using Eq. (4.9) with the comments following Eq. (4.9) taken into account. The resulting system is

$$
\begin{bmatrix}
2 & 1 & 0 & 0 & 0 \\
\frac{1}{2} & 2 & \frac{1}{2} & 0 & 0 \\
0 & \frac{1}{2} & 2 & \frac{1}{2} & 0 \\
0 & 0 & \frac{1}{2} & 2 & \frac{1}{2} \\
0 & 0 & 0 & 1 & 2
\end{bmatrix}
\begin{bmatrix}
0 \\ m_1 \\ m_2 \\ m_3 \\ 0
\end{bmatrix}
$$

$$
= 3
\begin{bmatrix}
-1/h & 1/h & 0 & 0 & 0 \\
-1/2h & 0 & 1/2h & 0 & 0 \\
0 & -1/2h & 0 & 1/2h & 0 \\
0 & 0 & -1/2h & 0 & 1/2h \\
0 & 0 & 0 & -1/h & 1/h
\end{bmatrix}
\begin{bmatrix}
0 \\ f_1 \\ f_2 \\ f_3 \\ 0
\end{bmatrix}.
\tag{4.11}
$$

If f_1, f_2, f_3 are chosen arbitrarily, the corresponding natural cubic spline will not have $m_0 = m_4 = 0$. The first and last rows of Eq. (4.11) immediately yield

$$
m_1 = 3f_1/h, \qquad m_3 = -3f_3/h,
\tag{4.12}
$$

whereas the remaining ones condense to the system

$$
\begin{bmatrix}
2 & \frac{1}{2} & 0 \\
\frac{1}{2} & 2 & \frac{1}{2} \\
0 & \frac{1}{2} & 2
\end{bmatrix}
\begin{bmatrix}
m_1 \\ m_2 \\ m_3
\end{bmatrix}
= \frac{3}{2h}
\begin{bmatrix}
f_2 \\ f_3 - f_1 \\ -f_2
\end{bmatrix}.
\tag{4.13}
$$

We must arrange f_1, f_2, f_3 so that the values m_1, m_3 coming from Eq. (4.13) coincide with those given in Eq. (4.12). On solving Eq. (4.13) for m_1 and m_3, we obtain

$$
m_1 = \frac{3}{14h}\left\{\frac{14}{4}f_2 + f_1 - f_3\right\}, \qquad m_3 = \frac{3}{14h}\left\{-\frac{14}{4}f_2 + f_1 - f_3\right\},
$$

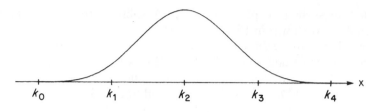

FIG. 4.5   *A B-spline.*

which must be equated to $3f_1/h$ and $-3f_3/h$, respectively. The resulting equations can be solved for f_1 and f_3 in terms of f_2 and give

$$f_1 = \tfrac{1}{4}f_2, \qquad f_3 = \tfrac{1}{4}f_2, \qquad f_2 \text{ non-zero, arbitrary.}$$

We could choose $f_2 = 1$; then, $f_1 = f_3 = \tfrac{1}{4}$. Thus, this minimal support B-spline $\hat{B}_0$ is defined by the ordinates $f_0 = f_4 = 0$, $f_1 = f_3 = \tfrac{1}{4}$, $f_2 = 1$, and the slopes $m_0 = m_4 = 0$, $m_1 = -m_3 - 3/4h$, $m_2 = 0$, and its values can be computed by using these in conjunction with the cardinal basis in the form of Eq. (3.12). A better choice of f_2 will be made shortly.

We are now in a position to construct all of the B-splines that are non-zero on the interval $[a, b] = [k_0, k_N]$ on which interpolation is being carried out. For simplicity we retain the hypothesis of equally spaced knots. First, we define the B-spline B_0, defined for all real numbers, by attaching identically zero extensions to $\hat{B}_0$:

$$B_0(x) = \begin{cases} 0, & -\infty < x < k_0, \\ \hat{B}_0(x), & k_0 \leq x \leq k_4, \\ 0, & k_4 < x < \infty. \end{cases} \qquad (4.14)$$

This is a piecewise cubic, degenerately so outside $[k_0, k_4]$, and has a continuous second derivative everywhere, because by definition the cubic spline $\hat{B}_0(x)$ is twice differentiable on (k_0, k_4). Furthermore, $\hat{B}_0''(k_0^+) = \hat{B}_0''(k_4^-) = 0$, and the second derivative is identically zero outside $[k_0, k_4]$. Now we define all of the B-splines whose support is entirely within $[k_0, k_N]$ by translating B_0 to the right:

$$B_j(x) = B_0(x - jh), \qquad j = 0, 1, \ldots, N - 4, \qquad k_j = jh + k_0. \qquad (4.15)$$

There are a few more B-splines whose support is not entirely in $[k_0, k_N]$ and which either "start" at the knots k_{-1}, k_{-2}, k_{-3} to the left of k_0, or terminate at the knots $k_{N+1}, k_{N+2}, k_{N+3}$ to the right of k_N (Fig. 4.6). Thus, the complete set of B-splines that have at least part of their support in $[k_0, k_N]$ are the restrictions to $[k_0, k_N]$ of the following translates of B_0:

$$B_j(x) = B_0(x - jh), \qquad j = -3, -2, -1, 0, 1, \ldots, N-1. \quad (4.16)$$

We see now that at any point of $[k_0, k_N]$, there are at most four B-splines that are not zero there (Fig. 4.6). In particular, consider any point x in $[k_i, k_{i+1}]$. The non-identically zero B-splines at this point are B_{i-3}, B_{i-2}, B_{i-1}, and B_i. It is interesting that the values of these four B-splines at x sum to a constant, which is independent of x, just as the tent functions sum to unity in the case of the linear spline.

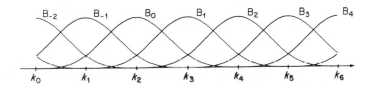

FIG. 4.6   B-splines.

Assuming the last assertion to be true, we can see what this sum must be by examining the sum at k_i (Fig. 4.6). We get

$$B_{i-3}(k_i) + B_{i-2}(k_i) + B_{i-1}(k_i) + B_i(k_i) = \tfrac{1}{4} + 1 + \tfrac{1}{4} + 0 = \tfrac{3}{2}.$$

We can get these splines to have a sum of unity by readjusting the value of f_2 in our construction of the minimal support prototype B-spline $\hat{B}_0$. What we want is

$$f_2 + \tfrac{1}{4}f_2 + \tfrac{1}{4}f_2 = 1;$$

from this,

$$f_2 = \tfrac{2}{3}, \qquad f_1 = f_3 = \tfrac{1}{6}$$

When this normalization is used, the following theorem can be proved. (Although our discussion has been in the context of equally spaced knots, the following results hold for unequal spacing of knots.)

Theorem 4.3.2   *Let the knots k_i, $i = -3, -2, \ldots, N+2, N+3$, satisfy $k_{i+1} > k_i$ and $k_0 = a$, $k_N = b$. Then, for any value x in $[a, b]$, the B-splines $B_{-3}, \ldots, B_{N-1}$ [Eq. (4.16)] satisfy*

$$\sum_{i=-3}^{N} B_i(x) = 1,$$

and any cubic spline $S(x)$ with the knots $k_0, \ldots, k_N$ can be written in the form

$$S(x) = \sum_{i=-3}^{N} a_i B_i(x). \tag{4.17}$$

There are $N + 3$ coefficients a_i in this representation, showing again that the vector space of cubic splines has dimension $N + 3$, so that the $N + 1$ function values will not determine $S(x)$ uniquely—two additional constraints must be supplied. Cardinality of the basis has been sacrificed for small support in the basis. Consequently, in evaluating $S(x)$ for any x in $[a, b]$, only four terms at most of the sum (4.17) will be non-zero.

4.4 Recursive construction of *B*-splines

The techniques developed so far do not lend themselves to the efficient calculation of a *B*-spline basis for $\mathcal{S}_N(K)$. They can be replaced by a recursive calculation, which we outline below. This technique offers a numerically stable computational process. Without striving for great generality (for which see de Boor [1978]), we first define *B*-splines that may be piecewise constant, linear, quadratic, or cubic polynomials.

Definition *Let* $k_i, i = -3, -2, \ldots, N + 3$, *be knots satisfying* $k_{-3} < k_{-2} < \ldots < k_{N+3}, k_0 = a, k_N = b$. *A B-spline of order n, n = 1, 2, 3, 4, with these knots is a piecewise $(n - 1)$th degree polynomial function not*

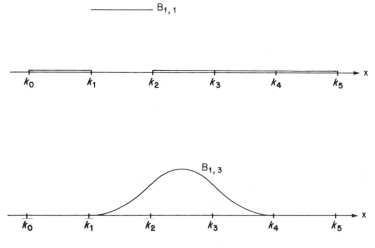

FIG. 4.7   *B*-splines of orders one and three.

identically zero, of continuity class $\mathscr{C}^{(n-2)}[k_{-3}, k_{N+3}]$, *and of minimal support. When* $n = 1$, *we interpret the class* $\mathscr{C}^{-1}[k_{-3}, k_{N+3}]$ *as admitting functions with discontinuities at the knots.*

We have already seen B-splines of order two, namely, the tent functions of Section 3.3, and of order four (although with equally spaced knots): These are our cubic B-splines. When $n = 1$ and $n = 3$, we have in mind piecewise constant and piecewise quadratic functions, as illustrated in Fig. 4.7. We may reasonably assume the existence of such functions. We can prove the following theorem as well.

Theorem 4.4.1 *The support of a B-spline of order n is* $n + 1$ *intervals. Denote by* $B_{i,n}$ *an nth-order B-spline whose support is* $[k_i, k_i + n]$ *(this contains* $n + 1$ *intervals created by the knot sequence). Then, it is possible to normalize these splines so that for any x in* $[a, b]$ *and* $n = 1, 2, 3, 4$,

$$\sum_{i=-3}^{N+3} B_{i,n}(x) = 1.$$

The cubic B-splines (of order $n = 4$) that we require can now be computed by means of Eq. (4.18), which is derived in de Boor (1978), and is used extensively in computation.

Theorem 4.4.2 *The B-splines of order n are related to those of order* $n - 1$ *by the recurrence relation*

$$B_{i,n}(x) = \frac{x - k_i}{k_{i+n-1} - k_i} B_{i,n-1}(x) + \frac{k_{i+n} - x}{k_{i+n} - k_{i+1}} B_{i+1,n-1}(x), \quad (4.18)$$

where $i = -3, \ldots, N - 1$ *and* $n = 1, 2, 3, 4$.

In computing with this relation we start with the B-spline of order one (Fig. 4.7). A precaution is necessary here in order to avoid uncertainty as to the value of, say, $B_{i,1}$ at k_i and k_{i+1}. A useful convention is to define the first-order splines as *right-continuous*, so that

$$B_{i,1}(k_i) = 1, \qquad B_{i,1}(k_{i+1}) = 0,$$

or, more generally,

$$B_{i,1} = \begin{cases} 0, & x < k_i, \\ 1, & k_i \leq x < k_{i+1}, \\ 0, & k_{i+1} \leq x. \end{cases} \qquad i = -3, -2, \ldots, N + 3, \quad (4.19)$$

The evaluation of $B_{i,4}(x)$ for any x in its support $[x_i, x_{i+4}]$ can be carried out by working within the following triangular array, working from left to right, one column at a time, according to Eq. (4.18):

$$
\begin{array}{llll}
B_{i,1} \\
B_{i+1,1} & B_{i,2} \\
B_{i+2,1} & B_{i+1,2} & B_{i,3} \\
B_{i+3,1} & B_{i+2,2} & B_{i+1,3} & B_{i,4}.
\end{array}
\qquad (4.20)
$$

For any x, the values in the first column will be 0 or 1, in view of Eq. (4.19).

As an illustration, consider the case of equally spaced knots and the associated triangular table for finding $B_{i,4}(k_j)$, $j = i + 1, i + 2$. For $x = k_{i+1}$ [cf. Eq. (4.19)]:

$$
\begin{array}{lllll}
B_{i,1}(k_{i+1}) & = 0 \\
B_{i+1,1}(k_{i+1}) & = 1 & 1 \\
B_{i+2,1}(k_{i+1}) & = 0 & 0 & \frac{1}{2} \\
B_{i+3,1}(k_{i+1}) & = 0 & 0 & 0 & \frac{1}{6}
\end{array}
$$

For $x = k_{i+2}$:

$$
\begin{array}{lllll}
B_{i,1}(k_{i+2}) & = 0 \\
B_{i+1,1}(k_{i+2}) & = 0 & 0 \\
B_{i+2,1}(k_{i+2}) & = 1 & 1 & \frac{1}{2} \\
B_{i+3,1}(k_{i+2}) & = 0 & 0 & \frac{1}{2} & \frac{2}{3}
\end{array}
$$

The values $\frac{1}{6}$ and $\frac{2}{3}$ are of course the same as those obtained by actual construction of $B_{i,4}$ earlier in this section. It is convenient to suppress the index 4 for simplicity of notation.

In order to construct an interpolating cubic spline in its cubic B-spline form [Eq. (4.17)], the conditions $S(k_j) = f_j$, $j = 0, 1, \ldots, N$, are imposed, giving

$$
\sum_{i=-3}^{N-1} a_i B_i(k_j) = f_j, \qquad j = 0, 1, \ldots, N. \qquad (4.21)
$$

The required values $B_i(k_j)$ can be computed via Eq. (4.18). In the jth equation, this is only necessary for the B-splines B_{j-3}, B_{j-2}, B_{j-1}, all other values being zero. Two more conditions have to be supplied as well, usually in the form of end conditions. For a *natural* spline these are $S''(k_0) = S''(k_N) = 0$, or in B-spline form,

$$a_{-3} B''_{-3}(k_0) + a_{-2} B''_{-2}(k_0) + a_{-1} B''_{-1}(k_0) = 0,$$

$$a_{N-3} B''_{N-3}(k_N) + a_{N-2} B''_{N-2}(k_N) + a_{N-1} B''_{N-1}(k_N) = 0. \tag{4.22}$$

The required second derivatives are given by

$$B''_{-3}(k_0) = \frac{6}{(k_1 - k_{-1})(k_1 - k_{-2})},$$

$$B''_{-2}(k_0) = \frac{-6}{k_1 - k_{-1}} \left\{ \frac{1}{k_2 - k_{-1}} + \frac{1}{k_1 - k_{-2}} \right\},$$

$$B''_{N-1}(k_0) = \frac{6}{(k_1 - k_{-1})(k_2 - k_{-1})},$$

$$B''_{N-3}(k_N) = \frac{6}{(k_{N+1} - k_{N-1})(k_{N+1} - k_{N-2})},$$

$$B''_{N-2}(k_N) = \frac{6}{k_{N+1} - k_{N-1}} \left\{ \frac{1}{k_{N+2} - k_{N-1}} + \frac{1}{k_{N+1} - k_{N-2}} \right\},$$

$$B''_{N-1}(k_N) = \frac{6}{(k_{N+1} - k_{N-1})(k_{N+2} - k_{N-1})}. \tag{4.23}$$

The extra knots $k_{-3}, k_{-2}, k_{-1}, k_{N+1}, k_{N+2}, k_{N+3}$ should satisfy $k_{-3} < k_{-2} < \ldots < k_{N+3}$, but are otherwise arbitrary. Once the a_i values are determined from Eqs. (4.21) and (4.22), the calculation of values of $S(x)$ for any $x \in [a, b]$ can be carried out by the use of the recurrence (4.18).

Clearly, the theory of B-splines is very involved, and we have only scratched the surface here. For problems with N not too large and which are not to be repeated many times with different parameter configurations, the computation of a piecewise polynomial representation for a cubic spline by means of the methods of Section 4.2 will prove satisfactory.

4.5 Natural cubic splines as optimal interpolants

A mechanical means of producing smooth interpolating curves, heavily used in the past in the shipbuilding and aircraft industries, consists in the use of a thin, flexible batten or spline, held in place by weights. This spline bends in such a way, that its internal energy due to bending is minimal, consistent with the interpolation constraints imposed on it. At any point of the spline, the bending energy depends on the curvature there. If the form of the interpolating spline is viewed as the graph of a function $S(x)$ with

$x \in [a, b]$, a and b being two of the points of interpolation, then it can be shown that the energy is proportional to the integral

$$\int_a^b \frac{[S''(x)]^2 \, dx}{[1 + (S'(x))^2]^{5/2}},$$

which can be rewritten as

$$\int_a^b [\kappa(x)]^2 [1 + (S'(x))^2]^{1/2} \, dx, \qquad (4.24)$$

where $\kappa(x)$ is the curvature of the curve $S(x)$ as defined in Section 1.4.

We can now pose the following problem: Find that function $S(x)$ which interpolates at the knots $a = x_0 < x_1 < \ldots < x_N = b$, and for which the integral of Eq. (4.24) exists and is least in value. By finding this minimizing function, we hope to reproduce mathematically the natural shape taken by the flexible spline. This is a rather difficult problem. A good discussion of it can be found in a paper by Malcolm (1977).

A simplification results if we ignore the contribution of the slope $S'(x)$ to the integral of Eq. (4.24) and set it to zero. Then the integral collapses to the functional [see the comments following Eq. (2.10)]

$$J(S) = \int_a^b [S''(x)]^2 \, dx. \qquad (4.25)$$

The set of all functions for which the above quantity can be computed is recognized as a useful mathematical entity. We denote it here by $\mathcal{I}_2[a, b]$ and refer to it as "the class of functions having a square integrable second derivative on $[a, b]$". The set $\mathcal{I}_2[a, b]$ contains, among others, all the functions of $\mathcal{P}_n$ for any n, all cubic splines on $[a, b]$, and all $\mathcal{C}^1$ piecewise cubics on $[a, b]$. The latter possess second derivatives everywhere except possibly at the knots. With an appropriate interpretation of the integral, this causes no difficulties.

Now we pose the problem: Find a function in $\mathcal{I}_2[a, b]$ that interpolates to given data at the knots $a = k_0 < k_1 < \ldots < k_N = b$ and minimizes the functional J of Eq. (4.25).

The solution is surprising. It is unique and is the *interpolating natural cubic spline*. Hence, even if the search for an optimal interpolant had been carried out in the smaller function classes $\mathcal{C}_N(K)$ (cf. Theorem 3.6.1) or $\mathcal{S}_N(K)$ (cf. Theorem 3.7.2), which are subspaces of $\mathcal{I}_2[a, b]$, the same solution would have been obtained. We can use this idea to develop natural cubic splines from the interpolating Hermitian piecewise cubics in $\mathcal{C}_N(K)$,

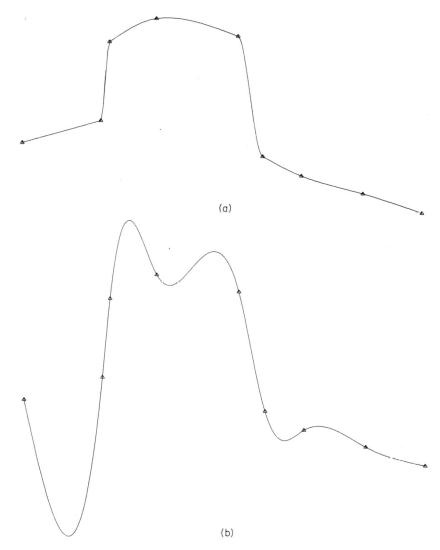

FIG. 4.8 (a) Weighted spline and (b) natural spline.

using the slopes as parameters, with respect to which the minimization of J can be carried out. The result is the set of Eq. (4.6), together with the natural end conditions.

The term "natural" refers to the fact that a mechanical spline that interpolates but is not subject to torques at the data point will flatten out to zero curvature at its extremities. This corresponds to zero second-derivative end conditions.

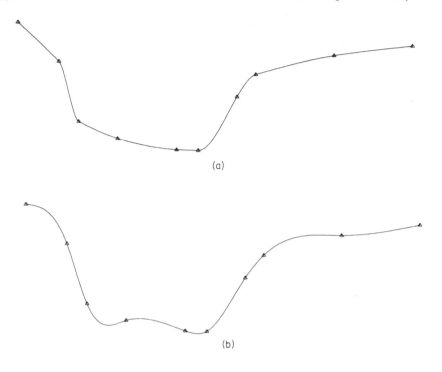

FIG. 4.9   (a) Weighted spline and (b) natural spline.

Although the natural cubic spline is optimal in the sense of minimizing Eq. (4.25), it should not be regarded as optimal in some universal sense. Different optimality criteria yield different optimal interpolants. An example is the weighted cubic spline in $\mathscr{C}_N(K)$, which interpolates and minimizes

$$\int_a^b w(x)\,[S''(x)]^2 \, dx,$$

where $w(x)$ is a positive weight function. A fairly elementary account of its theory can be found in Šalkauskas (1984). If $w(x)$ is constant, then a natural spline results. Otherwise, the spline is only of continuity class $\mathscr{C}^1$. The use of a suitable weight function allows this spline to accommodate itself to rapidly varying data without experiencing the severe over- and undershoot phenomena that occur often with $\mathscr{C}^2$ splines. Its defining equations (which yield the necessary slopes) are identical in appearance to those of the natural spline, but the values of the λ_i, μ_i, and c_i terms are computed somewhat differently.

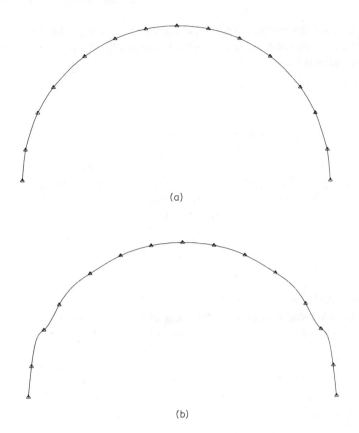

(a)

(b)

FIG. 4.10   (a) Weighted spline and (b) natural spline.

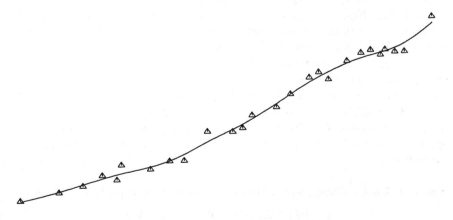

FIG. 4.11   Smoothing cubic spline; all weights 10.

Figures 4.8–4.10 compare the effects of one weighting scheme on various sets of data. Here we have used a simple weight function. It is a piecewise constant defined by

$$w(x) = \{1 + [(y_i - y_{i-1})/h_i]^2\}^{-3}, \qquad x_{i-1} \leqslant x < x_i, \quad i = 1, \ldots, N.$$

4.6 Smoothing with cubic splines: introduction

The idea of fitting a polynomial curve of degree m to more than $m + 1$ data points has been introduced in Section 2.5. The notion of fitting used there was a fit in the least squares sense. Thus, a polynomial p in the class $\mathcal{P}_m$ ($m + 1 \leqslant N$) was found so that the functional

$$E(p) = \sum_{j=0}^{N} (p(x_j) - f_j)^2 \tag{4.26}$$

was as small as possible.

It was noted that if we allowed $m + 1 = N$, the best least squares polynomial became capable of interpolating and thus reducing $E(p)$ to zero. On the other hand, we have another notion of smoothness associated with the functional J of Eq. (4.25). Whatever the magnitude of J may be for a particular interpolating spline S [which has $E(S) = 0$], it is obvious that J could likely be decreased if the condition $E(S) = 0$ were relaxed and interpolation were not required. This is the reasoning behind the creation of the functional

$$K_\lambda(f) = J(f) + \lambda E(f), \qquad \lambda > 0, \tag{4.27}$$

which is defined for all $f \in \mathcal{I}_2[a, b]$ and has the non-negative number λ as a parameter. Note that the data occurs in $E(f)$ but not in $J(f)$. Hence, if we allowed $\lambda = 0$, then $K_\lambda(f)$ would be minimized by *any* straight line, for a straight line has zero second derivative and hence zero $J(f)$. If λ is very close to zero, the minimizer of $K_\lambda(f)$ should be close to a least squares straight-line approximation to the data. On the other hand, if λ is very large, a minimization of $K_\lambda(f)$ would be dominated by $\lambda E(f)$, which is least when f interpolates. It turns out that the problem of minimizing $K_\lambda(f)$ for a fixed λ over all of $\mathcal{I}_2[a, b]$ (which contains many more functions than just splines) has a solution paralleling that of the minimization problems of Section 4.5.

Theorem 4.6.1   *There exists a unique function S_λ in $\mathcal{I}_2[a, b]$ for which*

$$\min_{f \in \mathcal{I}_2} \{J(f) + \lambda E(f)\} = J(S_\lambda) + \lambda E(S_\lambda),$$

i.e. which minimizes $K_\lambda(f)$, *and this minimizing function is a natural cubic spline on* $[a, b]$ *with knots* $a = x_0 < x_1 < \ldots < x_N = b$.

We note that the points $x_0, x_1, \ldots, x_n$ are fixed *a priori* and are used in the definition of $E(p)$.

Armed with this result, we may search in the small $(N + 1)$-dimensional subspace $\mathscr{S}_N(K)$ of $\mathscr{S}_2[a, b]$ composed of natural cubic splines with knot sequence $K = \{k_0, k_1, \ldots, k_N\}$ defined by $k_i = x_i$, $i = 0, 1, \ldots, N$. This permits a constructive description of S_λ. The only parameters at our disposal for the optimization of $K_\lambda(f)$ are just the ordinates y_j, $j = 0, 1, \ldots, N$, of a non-interpolating natural cubic spline at the knots $k_0 < k_1 < \ldots < k_N$. The required smoothing spline can be obtained by a simplified version of the method described next in Section 4.7.

4.7 The smoothing spline

A useful and more general functional than $K_\lambda(f)$ is obtained by modifying $E(f)$ in Eq. (4.26) by means of the provision of weights for the squares of the differences between the values of the approximating function f and the observations f_j. Hence we take

$$F_\lambda(f) = \sum_{j=0}^{N} \lambda_j (f(k_j) - f_j)^2, \tag{4.28}$$

with $\boldsymbol{\lambda} = [\lambda_0, \ldots, \lambda_N]^T$, and put

$$K_\lambda(f) = J(f) + E_\lambda(f). \tag{4.29}$$

Concerning the choice of $\boldsymbol{\lambda}$, it is evident that a large value of some λ_j in $\boldsymbol{\lambda}$ will, through the minimization process, cause $f(k_j)$ to be close to f_j, and near-interpolation of k_j will result. Thus, from a statistical viewpoint, the values λ_j should be inversely proportional to the variances σ_j^2 of the observations f_j. Hence we take

$$\lambda_j = \lambda/\sigma_j^2 \qquad j = 0, \ldots, N; \quad \lambda > 0. \tag{4.30}$$

Some statistical guidance is required here. We shall return to this question later in this section and concentrate on the problem of minimizing $K_\lambda(f)$ for $f \in \mathscr{S}_2[a, b]$. It turns out (Schoenberg, 1964) that the minimizing function is unique and is again a natural cubic spline with knots $a = x_0 < x_1 < \ldots < x_N = b$, i.e. Theorem 4.6.1 applies with a small change in notation.

In order to carry out the computation of the smoothing spline, it is tempting to attack the problem directly, working in terms of the yet unknown ordinates of the spline. This approach leads to complications that are avoided by a somewhat different strategy. The first move is to represent a natural spline by means of its ordinates and second, rather than first derivatives at the knots.

Let the values of the second derivatives at the knots be denoted by M_i, $i = 0, \ldots, N$. Then, since a spline $S(x)$ is in $\mathscr{C}^2[a, b]$, its second derivative is a continuous piecewise linear function in $\mathscr{L}_N(K)$ (Section 3.2), given explicitly by

$$S''(x) = M_{j-1}[(x_j - x)/h_j] + M_j[(x - x_{j-1})/h_j], \qquad x \in [x_{j-1}, x_j]. \tag{4.31}$$

Integrating twice, we obtain on $[k_{j-1}, k_j]$,

$$S(x) = \tfrac{1}{6}M_{j-1}[(x_j - x)^3/h_{j-1}] + \tfrac{1}{6}M_j[(x - x_{j-1})^3/h_j] + ax + b,$$

with integration constants a and b that can be evaluated by making sure that this segment of $S(x)$ interpolates to values y_{j-1} and y_j at the knots k_{j-1} and k_j, respectively.

Setting $S(k_{j-1}) = y_{j-1}$ and $S(k_j) = y_j$ and solving for a and b, we arrive at

$$\begin{aligned}
S(x) = {} & \tfrac{1}{6}M_{j-1}[(x_j - x)^3/h_j] + \tfrac{1}{6}M_j[(x - x_{j-1})^3/h_j] \\
& + \tfrac{1}{6}(6y_{j-1} - M_{j-1}h_j^2)[(x_j - x)/h_j] \\
& + \tfrac{1}{6}(6y_j - M_j h_j^2)[(x - x_{j-1})/h_j],
\end{aligned} \tag{4.32}$$

for any x in $[k_{j-1}, k_j]$.

Now we are assured of the continuity of $S''(x)$ as well as that of $S(x)$ (by interpolation), and we must make certain that the particular values of M_i that we are using permit $S'(x)$ to be continuous too. To achieve this goal, we compute $S'(x)$ for the interval $[k_{j-1}, k_j]$ and for its neighbouring interval $[k_j, k_{j+1}]$, evaluate it at $x = k_j$, and equate the two expressions to obtain Eq. (4.33):

$$\begin{aligned}
\tfrac{1}{6}h_j M_{j-1} &+ \tfrac{1}{3}(h_j + h_{j+1})M_j + \tfrac{1}{6}h_{j+1}M_{j+1} \\
&= (1/h_{j+1})(y_{j+1} - y_j) - (1/h_j)(y_j - y_{j-1}),
\end{aligned} \tag{4.33}$$

for $j = 1, \ldots, N - 1$. These are $N - 1$ equations relating the $N + 1$ second derivatives of $S(x)$ at the knots. As before, the two remaining degrees of freedom can be absorbed by imposing end conditions. It is traditional to divide Eq. (4.33) by $\tfrac{1}{6}(h_j + h_{j+1})$, $j = 1, \ldots, N - 1$, thereby arriving at a coefficient matrix very similar to that of Eq. (4.6). [The interested reader can find the details in Ahlberg et al. (1967).]

For our present purpose, the equations are best left as they are. Since we are looking for a *natural* cubic spline minimizing $K_\lambda(S)$, we can put $M_0 = M_N = 0$ at the outset and write Eq. (4.33) in the form

$$
\begin{bmatrix}
\frac{1}{3}(h_1 + h_2) & \frac{1}{6}h_2 & 0 & \cdots & & 0 \\
\frac{1}{6}h_2 & \frac{1}{3}(h_2 + h_3) & \frac{1}{6}h_3 & \cdots & & 0 \\
0 & \ddots & \ddots & & & \vdots \\
\vdots & & & \ddots & & \frac{1}{6}h_{N-1} \\
0 & \cdots & & 0 & \frac{1}{6}h_{N-1} & \frac{1}{3}(h_{N-1} + h_N)
\end{bmatrix}
\begin{bmatrix}
M_1 \\
M_2 \\
\vdots \\
\\
M_{N-1}
\end{bmatrix}
$$

$$(4.34)$$

$$
=
\begin{bmatrix}
\frac{1}{h_1} & -\frac{1}{h_1} - \frac{1}{h_2} & \frac{1}{h_2} & 0 & \cdots & & 0 \\
0 & \frac{1}{h_2} & -\frac{1}{h_2} - \frac{1}{h_3} & \frac{1}{h_3} & & & \vdots \\
\vdots & & \ddots & \ddots & \ddots & & \\
& & & & \frac{1}{h_{N-1}} & & 0 \\
0 & \cdots & & 0 & \frac{1}{h_{N-1}} & -\frac{1}{h_{N-1}} - \frac{1}{h_N} & \frac{1}{h_N}
\end{bmatrix}
\begin{bmatrix}
y_0 \\
y_1 \\
\vdots \\
\\
y_N
\end{bmatrix}
$$

or, more compactly, as

$$BM = Dy, \qquad\qquad (4.35)$$

noting that B is $(N-1) \times (N-1)$, D is $(N-1) \times (N+1)$, $\mathbf{M}$ is of length $(N-1)$, and $\mathbf{y}$ is of length $N+1$. Furthermore, B is a symmetric matrix that is diagonally dominant and consequently possesses a symmetric inverse.

We are now in a position to carry out the minimization of $K_\lambda(S)$. This procedure is outlined next and results in the linear systems (4.39) and (4.40), which are solved in turn for $\mathbf{M}$ and $\mathbf{y}$. In matrix–vector form,

$$
K_\lambda(S) = \int_a^b \mathbf{M}\mathbf{l}(x)\mathbf{l}(x)^T\mathbf{M}\, dx + (\mathbf{y} - \mathbf{f})^T\Lambda(\mathbf{y} - \mathbf{f})
$$

$$
= \mathbf{y}^T D^T (B^{-1})^T \int_a^b \mathbf{l}(x)\mathbf{l}^T(x)\, dx\, B^{-1}D\mathbf{y} + (\mathbf{y} - \mathbf{f})^T\Lambda(\mathbf{yf}).
$$

$$(4.36)$$

Here, $\mathbf{l}(x) = [l_1(x), \ldots, l_{N-1}(x)]^T$ is a vector of those tent functions that are needed to form $S''(x)$ for a natural spline, M_0 and M_N being zero, and Λ is the diagonal matrix

$$\Lambda = \text{diag}[\lambda_0, \ldots, \lambda_N].$$

When $K_\lambda(S)$ is differentiated with respect to $y_0, \ldots, y_N$ in order to obtain the necessary conditions for a minimum, we obtain the system

$$D^T(B^{-1})^T \int_a^b \mathbf{l}(x)\mathbf{l}^T(x)\, dx\, B^{-1}D\mathbf{y} + \Lambda(\mathbf{y} - \mathbf{f}) = \mathbf{0}. \tag{4.37}$$

The integral appearing here is a square matrix whose elements are

$$\int_a^b l_i(x)\, l_j(x)\, dx.$$

The computation of these is quite easy, and it turns out that this matrix is simply B. As a result, Eq. (4.37) simplifies to

$$D^T B^{-1} D\mathbf{y} + \Lambda(\mathbf{y} - \mathbf{f}) = \mathbf{0}. \tag{4.38}$$

Rather than solving this system for $\mathbf{y}$ and then solving Eq. (4.34) for $\mathbf{M}$ so as to be able to use Eq. (4.32) for the calculation of $S(x)$, we rewrite Eq. (4.38) in terms of $\mathbf{M}$ by using Eq. (4.35). We do this by first multiplying Eq. (4.38) on the left by $D\Lambda^{-1}$ and then replacing $D\mathbf{y}$ by $B\mathbf{M}$. Then,

$$(D\Lambda^{-1}D^T + B)\mathbf{M} = D\mathbf{f}, \tag{4.39}$$

where $\Lambda^{-1} = \text{diag}[1/\lambda_0, \ldots, 1/\lambda_N]$. This system can be solved for $\mathbf{M}$; then $\mathbf{y}$ can be computed from Eq. (4.38) by replacing $D\mathbf{y}$ in that equation by $B\mathbf{M}$ and rearranging it to obtain

$$\mathbf{y} = \mathbf{f} - \Lambda^{-1}D^T\mathbf{M}. \tag{4.40}$$

Equation (4.32) can now be used to compute $S(x)$ segment by segment.

We observe at this point that if all λ_i's are chosen large, then roughly speaking, Λ^{-1} is close to zero, the system of Eq. 4.39 is close to that in Eq. 4.35, and Eq. (4.40) is close to $\mathbf{y} = \mathbf{f}$. In other words, we are then close to interpolating with a natural cubic spline.

In order to obtain the solution S_λ of the simpler smoothing problem described in Section 4.6, it is only necessary to replace the matrix Λ by the scalar λ and, of course, Λ^{-1} by λ^{-1}.

Concerning the choice of $\boldsymbol{\lambda}$ and of the weights λ_i in general in a statistical setting, we refer the reader to the work of Craven and Wahba (1977) and Wahba (1981), where the ideas of cross-validation are discussed. In the simplest case, we can assume that all λ_i are equal, as in Section 4.6, and experiment with the value of λ until a satisfactory fit is obtained.

Figure 4.11 shows a smoothing approximation to the same velocity data as those used in Fig. 3.3. Other interesting illustrations and discussion can be found in de Boor (1978).

5

Curve Fitting and Projectors

5.1 The projector defined by simple polynomial interpolation

The fundamental theorem described in Section 2.1 tells us that, if we are given a set of $N + 1$ distinct points in an interval $[a, b]$ at which to interpolate, then we can associate with any function f in the class $\mathscr{C}[a, b]$ a *unique* interpolating polynomial p of the class $\mathscr{P}_N$. If the points of interpolation are $x_0, x_1, \ldots, x_N$, then this association is determined by $p(x_j) = f(x_j)$ for $j = 0, 1, \ldots, N$. We now use this association to define a new function, which we call P, defined on the set $\mathscr{C}[a, b]$ and with values in $\mathscr{P}_N$ (which can be seen as a subset of $\mathscr{C}[a, b]$). Thus, for any $f \in \mathscr{C}[a, b]$, $Pf = p$, where p is the polynomial interpolant for f in $\mathscr{P}_N$. [In this context it is customary to write Pf rather than $P(f)$.]

We recall that this function can be determined explicitly in the following way: If $p(x) = \sum_{j=0}^{N} a_j x^j$, then, as in Eqs. (2.1) and (2.2), the coefficients $a_0, \ldots, a_N$ are the components of the vector $\mathbf{a}$ given by

$$\mathbf{a} = V^{-1}\mathbf{f}. \tag{5.1}$$

Also, the components of $\mathbf{f}$ are $f(x_0), \ldots, f(x_N)$, and V is the Vandermonde matrix of the powers $1, x, x^2, \ldots, x^N$ evaluated at $x_0, x_1, \ldots, x_N$.

Because $\mathscr{P}_N$ is contained in $\mathscr{C}[a, b]$, we can apply the function P once more to the function $Pf = p$. Writing P^2 for the composition of P with itself, we get $P^2f = Pp$. However, the definition of P means that Pp is the polynomial interpolant for p, which is obviously just p itself, i.e. $Pp = p = Pf$. Thus, for any $f \in \mathscr{C}[a, b]$, $P^2f = Pf$, and this means that $P^2 = P$ (the effect of applying P twice is just the same as applying P once). This is the essential property of a "projector", and the function P described here is known as the "interpolating projector" for polynomial interpolation on the points $x_0, x_1, \ldots, x_N$. We say that P projects (or maps) the functions of $\mathscr{C}[a, b]$ onto $\mathscr{P}_N$, the *image* of P.

Consider now the functions in $\mathscr{C}[a, b]$, which are projected by P onto the identically zero function (which is, of course, a member of $\mathscr{P}_N$). Thus, we ask which functions $f \in \mathscr{C}[a, b]$ are such that $Pf = 0$. It is apparent from the definition of P that this is the case if and only if $f(x_0) = f(x_1) = \ldots = f(x_N) = 0$. The class of all such functions in $\mathscr{C}[a, b]$ is known as the *nullspace* (or kernel) of P; call it $\mathscr{H}$. For any function f in $\mathscr{C}[a, b]$, write

$$f = Pf + (f - Pf) = Pf + (I - P)f, \tag{5.2}$$

where I is the identity transformation; it maps any member of $\mathscr{C}[a, b]$ into itself. Since f and Pf have the same values at $x_0, x_1, \ldots, x_N$, the difference $f - Pf = (I - P)f$ must belong to $\mathscr{H}$. Thus, Eq. (5.2) tells us that any f in $\mathscr{C}[a, b]$ can be expressed as the sum of two functions—one of them (Pf) in $\mathscr{P}_N$ and the other $[(I - P)f]$ in $\mathscr{H}$. Also, it is not difficult to see that these components of f in $\mathscr{P}_N$ and $\mathscr{H}$ are unique. Referring to the heuristic diagram in Fig. 5.1 we want to develop the geometrical idea of the whole space

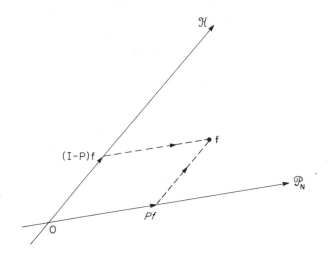

FIG. 5.1   Decomposition of $\mathscr{C}[a, b]$.

$\mathscr{C}[a, b]$ as a direct sum of $\mathscr{P}_N$ and $\mathscr{H}$ (written $\mathscr{C}[a, b] = \mathscr{P}_N \oplus \mathscr{H}$) and the component parts of a member f of $\mathscr{C}[a, b]$ obtained by projection onto $\mathscr{P}_N$ and onto $\mathscr{H}$.

Note that since $P^2 = P$, $I - P$ is also a projector for

$$(I - P)^2 = (I - P)(I - P) = I - 2P + P^2 = I - P.$$

Furthermore, $I - P$ projects the functions of $\mathscr{C}[a, b]$ onto $\mathscr{H}$.

5.2 A formal definition and further examples

In general, a function F defined on a vector space $\mathscr{S}$ ($\mathscr{S} = \mathscr{C}[a, b]$ in the above example) with values in a vector space $\mathscr{T}$ is said to be linear if it is *additive* and *homogeneous*; that is to say,

$$F(x + y) = F(x) + F(y),$$

for all x, y in $\mathscr{S}$, and

$$F(\alpha x) = \alpha F(x)$$

for any real α and all x in $\mathscr{S}$. Such a function is said to form a *linear transformation* from $\mathscr{S}$ into $\mathscr{T}$. It is often convenient to express this in the form $F : \mathscr{S} \to \mathscr{T}$.

It is easily verified that the projector introduced in Section 5.1 is a linear transformation from $\mathscr{C}[a, b]$ into $\mathscr{C}[a, b]$. In general, a *projector* is defined to be any linear transformation $P : \mathscr{S} \to \mathscr{S}$ with the property that $P^2 = P$. Such a linear transformation determines a decomposition of S as a direct sum of two subspaces, the *image* of P (onto which the members of $\mathscr{S}$ are projected) and the *nullspace* of P (consisting of members f of $\mathscr{S}$ annihilated by P, i.e, for which $Pf = 0$) (see Fig. 5.2)

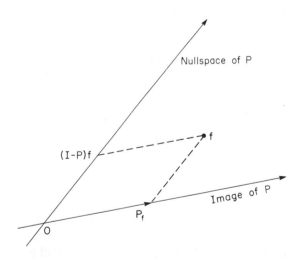

FIG. 5.2   Decomposition of $\mathscr{S}$ by a projector.

To understand fully a projector arising in a specific application, it is generally useful to look for a complete description of the two component subspaces: the image and the nullspace of P. Then the use of projectors

allows us to take a broader perspective of curve- and surface-fitting procedures, to rise above the computational detail, and to see more readily
how more complicated processes involving more than one projector can be
developed.

Let us illustrate with more examples. First, consider the case of Hermitian
polynomial interpolation, as described in Section 2.4. Theorem 2.4.1 shows
that, given $N + 1$ distinct points $x_0, x_1, \ldots, x_N$ in $[a, b]$ and any function f,
which, together with its first derivative, is well defined at these points, there
is a unique associated polynomial p in $\mathscr{P}_{2N+1}$ that takes the same function
and derivative values as f at $x_0, x_1, \ldots, x_N$. In particular, we may take any
f in $\mathscr{C}^1[a, b]$ and map it in this way onto a p in $\mathscr{P}_{2N+1}$. This map determines
the function H, say. Thus, we write $Hf = p$.

Because of the uniqueness of the Hermitian interpolant, the function H
will simply reproduce functions p in $\mathscr{P}_{2N+1}$. Thus, if $Hf = p$, then $H^2f =
Hp = p$, and $H^2f = Hf$ for any f in $\mathscr{C}^1[a, b]$. Again, H is linear, and so it is
a projector of $\mathscr{C}^1[a, b]$ onto the subspace $\mathscr{P}_{2N+1}$ of $\mathscr{C}^1[a, b]$. In this case,
the nullspace of H obviously consists of those functions f in $\mathscr{C}^1[a, b]$ for
which

$$f(x_0) = f(x_1) = \ldots = f(x_N) = 0, \qquad f^{(1)}(x_0) = f^{(1)}(x_1) = \ldots = f^{(1)}(x_N) = 0.$$

We turn now to the method of least squares, as described in Section 2.5.
Again, we may start with a function f from $\mathscr{C}[a, b]$, which is evaluated at
$N + 1$ points, $x_0, x_1, \ldots, x_N$. Then, if m is an integer, with $0 \leqslant m \leqslant N$, the
least squares technique described there associates a unique polynomial
$p \in \mathscr{P}_m$ with f, i.e. the technique determines a function $L : \mathscr{C}[a, b] \to \mathscr{P}_m$.
This association is given computationally by Eq. (2.14), i.e.

$$\mathbf{a} = (V^T V)^{-1} V^T \mathbf{f}, \tag{5.3}$$

where $p(x) = \sum_{j=0}^m a_j x^j$, $\mathbf{f}^T = [f(x_0), \ldots, f(x_N)]$ and V is the Vandermonde
matrix of $1, x, \ldots, x^m$ at $x_0, x_1, \ldots, x_N$.

Again, we easily see that if $p \in \mathscr{P}_m$, then $Lp = p$ and hence $L^2 = L$.
Also, L is linear and is therefore a projector of $\mathscr{C}[a, b]$ onto $\mathscr{P}_m$. The null
space of L is, however, harder to describe. It certainly contains all those
functions in $\mathscr{C}[a, b]$ that vanish at $x_0, x_1, \ldots, x_N$, but will generally contain
more.

There is a projector associated with the interpolating moving least squares
technique of Section 2.10, but this is, as one might expect, rather more
complicated. If, as above, there are $N + 1$ data points in $[a, b]$ and m is an
integer with $0 \leqslant m \leqslant N$, then it can be shown that the associated projector

defined on $\mathscr{C}[a, b]$ has an image of dimension $N + 1$. Indeed, a basis for this image can be made up of the $N + 1$ cardinal functions obtained by setting

$$f(x_j) = 1 \quad \text{and} \quad f(x_i) = 0 \quad \text{for} \quad i \neq j,$$

and $j = 0, 1, \ldots, N$. It can also be shown that $\mathscr{P}_m$ is contained in this image. The functions in the image of the projector are no longer mere polynomials. *At each* x, the value of the image of f is given by solving Eq. (2.26) and then setting $(Pf)(x) = \Sigma_{j=0}^{m} a_j x^j$. Thus, the coefficients a_j also depend on x. [For more details, the interested reader is referred to Lancaster and Šalkauskas (1981).]

Turning now to piecewise polynomial interpolants, we consider interpolation by linear splines, as described in Section 3.3. The vector space $\mathscr{L}_N(K)$ of linear splines with knot sequence K is a subspace of $\mathscr{C}[a, b]$, and every member of $\mathscr{L}_N(K)$ can be expressed as a linear combination of the functions $|x - k_i|, i = 0, 1, \ldots, N$. For any $f \in \mathscr{C}[a, b]$, there exists a unique linear spline l in $\mathscr{L}_N(K)$ interpolating f at the knots in K. This spline depends linearly on the values of f. Thus, there is defined a linear transformation L such that $Lf = l$. Furthermore, it is clear that $Ll = l$, so that $L^2 = L$, and L is a projector.

The representation of the function Lf depends on our choice of basis for $\mathscr{L}_N(K)$. If we choose the set $|x - k_i|, i = 0, 1, \ldots, N$, then $Lf(x) = \Sigma_{i=0}^{N} a_i |x - k_i|$. Here the coefficients a_i are given by

$$\mathbf{a} = V^{-1}\mathbf{f}, \tag{5.4}$$

where V is the Vandermondian of the basis shown in Eq. (3.1). On the other hand, if we choose a cardinal basis of tent functions, then the associated Vandermondian is simply the identity matrix I, and the matrix inversion indicated in Eq. (5.4) is avoided. The same applies to any interpolation method; however, it may be difficult to know ahead of time what the cardinal functions are.

To conclude this section, we consider the following fairly general problem. Let $\Phi_0, \Phi_1, \ldots, \Phi_N$ be linearly independent functions defined and at least continuous on $[a, b]$. Then the set of all linear combinations $\Sigma_{i=0}^{N} a_i \Phi_i$ forms a subspace $\mathscr{T}$ of $\mathscr{S} = \mathscr{C}[a, b]$. Suppose now that we desire an interpolant $\Phi \in \mathscr{T}$ to a function f in $\mathscr{C}[a, b]$ interpolating at distinct points $x_0, x_1, \ldots, x_N$ in $[a, b]$. Then, given the existence of Φ, it follows that $\Phi(x_j) = f(x_j), j = 0, 1, \ldots, N$, and since Φ is a linear combination of $\Phi_0, \Phi_1, \ldots, \Phi_N$,

$$\sum_{i=0}^{N} a_i \Phi_i(x_j) = f(x_j), \quad j = 0, 1, \ldots, N,$$

for some numbers $a_0, a_1, \ldots, a_N$. This is, in matrix form,

$$Va = f. \tag{5.5}$$

Here, V is the Vandermondian $[\Phi_i(x_j)]$. This matrix is also called a generalized Grammian. Its invertibility is related to the idea of unisolvency. [For a good discussion of this and other matters related to interpolation, see Davis (1963).] If V is invertible, a unique interpolant Φ does indeed exist and can be written in the form

$$\Phi = [\Phi_0, \ldots, \Phi_N] \begin{bmatrix} a_0 \\ \vdots \\ a_N \end{bmatrix} = \Phi^T a = \Phi^T V^{-1} f. \tag{5.6}$$

The row vector of functions formed by the product $\Phi^T V^{-1}$ is in fact a vector of cardinal functions, say $\psi^T = [\psi_0, \ldots, \psi_N]$. The reason for this is that the row vector Φ^T becomes a row of V when $x = x_i$, $i = 0, 1, \ldots, N$, and hence $\Phi^T V^{-1}$ becomes a row of the identity matrix, revealing that all ψ_j's are zero at x_i, except for ψ_i, which has value 1 there. We deduce that every interpolating projector $P : \mathcal{S} \to \mathcal{T}$ of the form considered here can be expressed in terms of cardinal functions:

$$Pf = \Phi = \Phi^T a = \psi^T f. \tag{5.7}$$

We shall find representations of this kind useful in the sequel.

5.3 A general least squares projector

This section contains a brief discussion of a rather general form of least squares projector, which contains the classical method of Section 2.5 as a special case and will be useful later. Let $\Phi_0, \Phi_1, \ldots, \Phi_m$ be linearly independent functions in $\mathscr{C}[a, b]$, and again let $\mathcal{T}$ be the vector space of all linear combinations $\sum_{i=0}^{m} a_i \Phi_i$. Assume that $m \le N$. Now, in general we cannot find a Φ in $\mathcal{T}$ that would interpolate a function $f \in \mathscr{C}[a, b]$ at $N + 1$ distinct points $x_0, x_1, \ldots, x_N$ of $[a, b]$ because the dimension of $\mathcal{T}$ is too small. The least squares problem analogous to that of Section 2.5 is to find a Φ to approximate f in the sense of minimizing

$$E(\Phi) = \sum_{i=0}^{N} [\Phi(x_i) - f(x_j)]^2$$

by a suitable choice of coefficients a_i involved in the formation of Φ. A more general functional is

$$E_W(\Phi) = \sum_{i=0}^{N} \sum_{j=0}^{N} [\Phi(x_i) - f(x_i)]w_{ij}[\Phi(x_j) - f(x_j)], \qquad (5.8)$$

into which some "weights" w_{ij} are introduced and which reduces to the functional of Section 2.5, when the matrix $W = [w_{ij}] = I$ and when $\Phi_0(x) = 1$, $\Phi_1(x) = x, \ldots, \Phi_m(x) = x^m$. Without loss of generality, we can assume that W is symmetric; that is, $w_{ij} = w_{ji}$, $i, j = 0, 1, \ldots, N$.

By computing the partial derivatives $\partial E_W(\Phi)/\partial a_i$, $i = 0, 1, \ldots, m$, and equating them to zero, we obtain the following *normal equations* for the $m + 1$ unknowns $a_0, \ldots, a_m$:

$$B^T W B\, \mathbf{a} = B^T W \mathbf{f}, \qquad (5.9)$$

in which

$$B = \begin{bmatrix} \Phi_0(x_0) & \Phi_1(x_0) & \ldots & \Phi_m(x_0) \\ & & \vdots & \\ \Phi_0(x_N) & \Phi_1(x_N) & \ldots & \Phi_m(x_N) \end{bmatrix}.$$

The matrix B is the analogue of the matrix V in Eq. (2.13).

Provided that the $(m + 1) \times (m + 1)$ matrix $B^T W B$ is invertible, we obtain

$$\mathbf{a} = (B^T W B)^{-1} B^T W \mathbf{f}, \qquad (5.10)$$

the analogue of Eq. (5.5) and see that $\mathbf{a}$ depends linearly on $\mathbf{f}$. Since the function Φ is given by $\Phi = \mathbf{\Phi}^T \mathbf{a}$ [see Eq. (5.7)], this relation determines a procedure for obtaining Φ from f. That is, we have defined a function Q_W whose defining property is $Q_W f = \Phi$. In fact, Q_W is a projector because $Q_W f = \Phi \in \mathcal{T}$ and $Q_W \Phi = \Phi$, and hence $Q_W^2 f = Q_W f$ for any $f \in \mathscr{C}[a, b]$. Also, Q_W will be an interpolating projector if we put $m = N$. If f is by chance already in $\mathcal{T}$, i.e. if it is representable by some linear combination of $\Phi_0, \Phi_1, \ldots, \Phi_m$, then $Q_W f = f$ and Eq. (5.10) will reveal the precise nature of the combination.

For $B^T W B$ to be nonsingular it is sufficient but not necessary for W to be positive definite. It turns out that the W employed next in Section 5.4 to construct *natural* cubic splines is indefinite. [See Bos and Šalkauskas (1987) for more details.]

5.4 Cubic spline projectors

Although it was quite simple to construct the interpolating projector that yields linear splines, this is not the case for $\mathscr{C}^2$ cubic splines. The difference stems from the fact that the dimension of the space $\mathscr{C}_N(K)$ of $\mathscr{C}^2$ cubics with knots $k_i = x_i$, $i = 0, 1, \ldots, N$, is $N + 3$ (cf. Section 3.7), which is two too many for unique interpolation at the knots, whereas the space $\mathscr{L}_N(K)$ of linear splines has just the correct dimension. On the other hand, there is an important feature common to both that will provide the mechanism for constructing the cubic spline projector.

Consider Eqs. (3.8) and (3.10) for linear and cubic splines respectively:

$$l(x) = Ax + B + \sum_{i=1}^{N-1} a_i |x - k_i|, \tag{5.11}$$

$$S(x) = \tfrac{1}{3}Ax^3 + \tfrac{1}{2}Bx^2 + Cx + D + \tfrac{1}{6} \sum_{i=1}^{N-1} a_i |x - k_i|^3. \tag{5.12}$$

We intend to interpolate on the interval $[a, b]$ with knots satisfying $a = k_0 < k_1 < \ldots < k_N = b$, and $x_i = k_i$, $i = 0, 1, \ldots, N$. Then we may also use the representations (cf. Section 3.2):

$$l(x) = \sum_{i=0}^{N} a_i |x - k_i|, \tag{5.13}$$

$$S(x) = ax + b + \tfrac{1}{6} \sum_{i=0}^{N} a_i |x - k_i|^3. \tag{5.14}$$

Here we have used the facts that $|x - k_0| = x - k_0$, when $x \geq k_0$, and $|x - k_N| = k_N - x_N$, when $x \leq k_N$, so that the terms $Ax + B$ from $l(x)$ and most of $\tfrac{1}{3}Ax^3 + \tfrac{1}{2}Bx^2 + Cx + D$ from $S(x)$ can be absorbed into the summations.

From Eq. (5.11) we see that $l(x)$ is capable of being a straight line, whereas Eq. (5.12) shows that $S(x)$ can be a simple cubic. The representation (5.14) shows that $S(x)$ has a part that is similar to $l(x)$ (the summation), with $(N + 1)$ degrees of freedom and a straight-line part $ax + b$ that contributes two more degrees of freedom. This suggests that a non-zero function of the form $ax + b$ cannot be represented as a sum $\sum_{i=0}^{N} a_i |x - k_i|^3$ for *any* choice of $a_0, a_1, \ldots, a_N$. Consequently, if we were to construct an interpolating projector P using only $N + 1$ basis functions $|x - k_i|^3$, $i = 0, 1, \ldots, N$, and the general method of Section 5.2, then Pf will not be able to be a straight line even if the data were to lie on a line.

This deficiency can be overcome by forming the following *composite projector*, known as a *Boolean sum*, out of P and *any* projector Q from $\mathscr{C}[a, b]$ onto $\mathscr{P}_1$. In fact, a least squares projector Q_W of Section 5.3, with $m = 1$ and $\Phi_0(x) = 1$, $\Phi_1(x) = x$, will do. Also, it should be realized that there is a lot of freedom in the choice of the symmetric matrix W. (Indeed, we will see that in some important cases, W can be indefinite in the sense that it has both positive and negative eigenvalues.) Our composite projector is written $P \oplus Q$ (the order is important) and is defined by its action on f:

$$(P \oplus Q)f = P(f - Qf) + Qf. \tag{5.15}$$

It is easy to see that $(P \oplus Q)f$ has the structure of Eq. (5.14), for Qf is a linear function $ax + b$, whereas $P(f - Qf)$ is a linear combination of the functions $|x - k_i|^3$, $i = 0, 1, \ldots, N$.

If f happens to have values at the knots that lie on a straight line, we will suppose that f *is* a line. Then in light of the comments of Section 5.3, $Qf = f$ and $(P \oplus Q)f = P(0) + f = f$, for the interpolant of the zero function (or of zero data) is zero. Hence the Boolean sum preserves those functions that are in the image of Q, $\mathscr{P}_1$ in this case. It also interpolates. To see this, notice that $P(f - Qf)$ must have values $f(k_i) - (Qf)(k_i)$ at the knots because P is an interpolating projector. Therefore $(P \oplus Q)f$ has values $f(k_i) - (Qf)(k_i) + (Qf)(k_i) = f(k_i)$. Using $Q = Q_W$ and the concrete representations of P and Q_W given by Eqs. (5.6) [with Eq. (5.7)] and (5.10), we can summarize these results as follows.

Theorem 5.6.1   *Let P be the interpolating projector formed from the basis functions $|x - k_i|^3$, $i = 0, 1, \ldots, N$, and let Q_W be a least squares projector onto $\mathscr{P}_1$. Then $S_W = P \oplus Q_W$ is an interpolating projector, and for any $f \in \mathscr{C}[a, b]$, $S_W f$ is a cubic spline. Furthermore, $S_W f$ has the explicit form*

$$
\begin{aligned}
S_W f &= P(f - Q_W f) + Q_W f \\
&= [|x - k_0|^3, |x - k_1|^3, \ldots, |x - k_N|^3] V^{-1} [\mathbf{f} - B(B^T W B)^{-1} B^T W \mathbf{f}] \\
&\quad + [1, x](B^T W B)^{-1} B^T W \mathbf{f},
\end{aligned}
$$

where

$$V = [|k_i - k_j|^3]_{i,j=0}^N \quad and \quad B^T = \begin{bmatrix} 1 & 1 & \cdots & 1 \\ k_0 & k_1 & \cdots & k_N \end{bmatrix}.$$

(It can be shown that the matrix V is non-singular so that $S_W f$ is well-defined whenever $B^T W B$ is also non-singular.)

Since the weight matrix W is as yet unspecified, the Boolean sum $P \oplus Q_W$ generates a large class of spline projectors. This formulation replaces the freedom to choose end conditions as described in Section 4.2 with the possibility of choosing weight matrices W. It turns out that the simple splines interpolating linear data but with derivative end conditions that are not consistent with the linear data cannot be produced in this way. However, the following important results can be proved:

Theorem 5.4.2   (a) *For any nonsingular, symmetric, weight matrix W, the class of functions f for which $S_W f = f$ (the image of S_W) has dimension $N + 1$.*
(b) *If $W = V^{-1}$, then $B^T W B$ is non-singular, and $S_W = P \oplus Q_W$ is the interpolating natural spline projector.*

We show next in Section 5.5 that the transformation S_W is indeed a projector. This result gives us an alternative description of the interpolant that is best in the sense of minimizing the measure of curvature or bending energy given by Eq. (4.25). As well, when formulated in this way, natural spline interpolation can be interpreted as a one-variable version of the interpolating process known in geostatistics as "kriging" with generalized covariance $|x|^3$. [See, for example, Delfiner (1975) and Huigbregts and Matheron (1970).]

It is possible to choose a much simpler projector Q onto $\mathcal{P}_1$. In particular, one could select any two knots and define Qf as the straight-line interpolant to f at these knots. Then the two necessary extra degrees of freedom that are not used up by the interpolation conditions must somehow be incorporated into the projector P in order for $P \oplus Q$ to represent an arbitrary interpolating cubic spline projector. This is essentially what happens in the process used by Meinguet (1978) for the construction of optimal interpolants.

5.5 Interpolation with Boolean sums of projectors

In Section 5.4 we met the Boolean sum of projectors in the context of cubic spline interpolation. Now we take a somewhat broader point of view and examine a wide class of interpolation schemes that include cubic splines as a special case.

The projectors we have discussed have been shown to act on functions f in $\mathscr{C}[a, b]$. In practical terms, however, our knowledge of a particular f is limited to some values of f at the points x_i, $i = 0, 1, \ldots, N$, in $[a, b]$. The projectors we employ use only these values. We then hope that Pf computed from them is a good representation of the function f underlying the data.

We also expect our projectors to recognize when the function values lie on a straight line or perhaps a parabola and in such a case to produce the linear or quadratic function, respectively. In symbols, then, we want P to be such that

$$Pf = f \qquad \text{for all} \quad f \quad \text{in} \quad \mathcal{P}_m, \tag{5.16}$$

and m is typically 0, 1, or 2.

We also want P to be an interpolating projector; it will be constructed much as in Section 5.2 from some basis functions $\Phi_0, \ldots, \Phi_N$. The shapes of these individual functions will influence the shape of the interpolant. If we choose P exactly as in Eq. (5.6), then most likely Eq. (5.16) will be violated. We can, however, achieve our aim by the use of a Boolean sum of projectors.

Theorem 5.5.1   *Let R be a projector from $\mathcal{C}[a, b]$ onto $\mathcal{P}_m$, and let Q be an interpolating projector defined on $\mathcal{C}[a, b]$ for interpolation at the distinct points $x_0, x_1, \ldots, x_N$ in $[a, b]$ (with $N \geq m$). Then the linear transformation $Q \oplus R$ (Boolean sum) defined on $\mathcal{C}[a, b]$ by $(Q \oplus R)f = Q(f - Rf) + Rf$ has the interpolation property $(Q \oplus R)f(x_i) = f(x_i)$, $i = 0, 1, \ldots, N$, and the property $(Q \oplus R)f = f$, $f \in \mathcal{P}_m$. In addition, if $RQ = R$, then $Q \oplus R$ is a projector.*

The meaning of $RQ = R$ is that $(RQ)f = Rf$ for all $f \in \mathcal{C}[a, b]$, and $(RQ)f = R(Qf)$. This always holds for our projectors because R needs only information about the values of f at the x_i's and since Qf is an interpolant, R "sees" the same values in Qf as in f.

The verification of the claims of Theorem 5.5.1 concerning interpolation and property (5.16) proceeds exactly as in Section 5.4. To show that $Q \oplus R$ is a projector, we write it in the form $Q - QR + R$ and "square" it carefully.

$$(Q - QR + R)(Q - QR + R)$$
$$= Q^2 - Q^2R + QR - QRQ + QRQR - QR^2 + RQ - RQR + R^2.$$

Since Q and R are projectors, $Q^2 = Q$ and $R^2 = R$. Using this together with $RQ = R$, we get

$$(Q \oplus R)^2 = Q - QR + R = Q \oplus R.$$

Since Q and R are linear, so is $Q \oplus R$, and hence it is a projector.

We have now verified that the transformation S_W of Section 5.4 is indeed a projector, as claimed in Theorem 5.4.1.

At this point it is worth summarizing the ingredients necessary for the formation of an interpolating projector P of Boolean sum type. We need the following:

(1) m, the degree of the polynomials preserved by P,
(2) R, a projector onto $\mathcal{P}_m$, and
(3) Q, an interpolating projector onto a space spanned by some functions $\Phi_0, \ldots, \Phi_N$.

The choice of m is not very critical; however, a value of $m = 1$ at least is recommended so that "flat" data will have a flat interpolant. In this case, R is any projector onto $\mathcal{P}_1$. For example, the general least squares projector as discussed at the end of Section 5.2 is suitable. For the time being, the weight matrix remains arbitrary—we shall have more to say about it shortly. In order to define Q, we must choose the functions $\Phi_0, \ldots, \Phi_N$. Since the Vandermondian matrix (see, e.g. Eq. (5.5) et seq.) of these functions comes into play in the construction of Q, some care must be taken to ensure its invertibility. It is known that the choice $\Phi_0(x) = 1$, $\Phi_1(x) = x, \ldots,$ $\Phi_N(x) = x^N$, results in a non-singular Vandermondian; however, this choice leads back to polynomial interpolants that are often unsuitable and, in any case, low-degree polynomial behaviour is accounted for by the projector R. We therefore do *not* choose polynomials for the Φ_i's.

A source of functions that have a non-singular Vandermondian can be found in the theory of Chebyshev systems and the theory of totally positive kernels [see Karlin and Studden (1966) and Karlin (1968)]. These theories are difficult and we content ourselves with some examples.

EXAMPLE 1   The functions $\Phi_0(x) = 1$, $\Phi_1(x) = e^x, \ldots, \Phi_N(x) = e^{Nx}$ have a non-singular Vandermondian (its determinant is strictly positive for all values of N).

EXAMPLE 2   The functions $\Phi_0(x) = e^{x_0 x}$, $\Phi_1(x) = e^{x_1 x}, \ldots, \Phi_N(x) = e^{x_N x}$ have a non-singular Vandermondian.

EXAMPLE 3   The functions $\Phi_0(x) = |x - x_0|^\lambda, \ldots, \Phi_N(x) = |x - x_N|^\lambda$, $\lambda = 1, 3, \ldots$, have a non-singular Vandermondian.

5.6 Concluding remarks on the definition of projectors Q and R

It is reasonable to seek interpolants that are translation-invariant in the sense that a horizontal shift in the data simply results in a shifted interpolant. Polynomial interpolation has this property, as do the interpolants constructed from the functions of Examples 1 and 3 above. More generally, one may attempt to employ translates of a fixed function Φ for the basis.

Thus,

$$\Phi_i(x) = \Phi(x - x_i), \qquad i = 0, 1, \ldots, N. \qquad (5.17)$$

Of course, Φ must be defined on a domain that is large enough to admit the definition of the translates Φ_i on $[x_0, x_n]$. If the resulting functions have a nonsingular Vandermondian, they can be used for the construction of an interpolating projector. It is also desirable that the interpolation process be independent of the direction of the x axis. To achieve this in conjunction with translation-invariance, one chooses Φ so that $\Phi(-x) = \Phi(x)$. Examples are $\Phi(x) = |x|$, which gives rise to linear splines, and $\Phi(x) = |x|^3$, which generates cubic splines.

Some attempts have been made to tailor the choice of Φ to the nature of the data. This has been done in a two-dimensional context and is discussed in Chapter 11 in connection with the interpolation process known as kriging.

We conclude with some remarks concerning the choice of the projector R that accounts for low-degree polynomial behaviour in the data. It seems reasonable to suppose that there is an optimal choice of R once m, the degree of polynomials to be preserved by the interpolation process, and the functions Φ_i, which are to account for a non-polynomial form of the interpolant, are selected.

In fact, if the Vandermondian of the functions Φ_i is invertible and has a certain additional property, then an optimal choice of R is that weighted least-squares approximant of the form discussed in Section 5.3 whose weight matrix is the inverse of the Vandermondian. The additional property to which we refer is known as *conditional positive-definiteness of order* $m + 1$ of the Vandermondian. It turns out that the Vandermondian of the functions $|x - x_i|^3$ has this property when $m = 1$ and that a corresponding optimal interpolant exists. It is in fact the natural cubic spline of Corollary 5.4.2, which is also optimal in the sense of minimizing the functional $J(s)$ of Eq. (4.25). In the present context, the optimality results from minimizing an upper bound on the error of interpolation.

Again, these ideas form a part of the kriging process. In contrast to the deterministic arguments used here, they are normally derived from statistical considerations and in a multivariate setting. We shall return to this topic in Chapter 11.

6

Criteria for Curve and Surface Fitting

We are now at a mid-point of our development; we have dealt with curve fitting, as far as we wish to go, and are about to embark on the problems of surface fitting. The sequence of techniques will be broken at this point to allow for reflection on the considerations that are likely to go into the choice of an appropriate technique for a particular problem. There are no universal answers here: The technique chosen should depend on the nature of the data, the nature of the phenomenon modelled (as far as it is known), and the characteristics of the technique considered by the user to be most important. We shall list some of these general characteristics and illustrate them by referring back to the techniques developed in Chapters 2–4. However, they will all apply (with small changes of language) to the discussion of surface fitting.

6.1 Differentiability of the fitted curve

By far the simplest curve-fitting procedure is the linear spline. That is, the points (x_i, f_i), (x_{i+1}, f_{i+1}) are joined by straight segments, giving rise to a curve of class $\mathscr{C}$. In general, this is not acceptable because it is known that the physical curve we are trying to represent is not like that. If the data points were sufficiently dense, then the broken linear curve may be acceptable, but this is not usually the case. Even so, it is worth bearing in mind that the piecewise linear interpolant does very well in the representation of ramps or steps, such as those which may well occur in sections of geological faults, for example.

Also, it is possible to refine successively a linear interpolant until it is apparently smooth. In the case of surface fitting, a similar phenomenon occurs (see Sections 7.4 and 7.9). This is the idea behind the algorithm of Ichida discussed in Section 3.4.

If the user agrees that piecewise linear techniques produce fits that really are not smooth enough, then the next natural question is: How smooth *should* the fitted curve be? For an answer to this question it is useful to think in terms of the classes of functions $\mathscr{C}^r$, $r = 0, 1, 2, \ldots$, introduced in Section 1.3. A linear spline is of class $\mathscr{C}$ and a curve with a well-defined tangent at every point (having no "corners") is of class $\mathscr{C}^1$. If the curve representing the *slope* of a $\mathscr{C}^1$ curve is itself of class $\mathscr{C}^1$, then the original curve is of class $\mathscr{C}^2$, and so on.

It is important to bear in mind what is meant by "smooth" in this context. We now say that a curve is "very smooth" on a set S if it has "many" successive derivatives at *every* point of S. We are thinking of smoothness in the sense of differentiability. Thus, in this sense, the polynomial curves of Fig. 2.10 are smoother than the spline curves of Fig. 4.4. The polynomial function has as many derivatives as we care to compute at *any* value of x. In contrast, there are points at which the spline has only two derivatives.

If a curve of smoothness (differentiability) class $\mathscr{C}^N$ is required, then, for example, one might use a spline technique of degree $N + 1$ or higher. If very high differentiability is required, then this may indicate the fit of a polynomial (all polynomials are of class $\mathscr{C}^\infty$), either by interpolation or by a least squares method. However, it is worth having in mind that the eye probably cannot distinguish among curves of the classes $\mathscr{C}^r$, where $r \geqslant 2$.

6.2 Confidence in the data

Does the user have sufficient confidence in his data to demand that the fitted curve must contain (i.e. interpolate) every data point? Or, is there a sizeable experimental error that enters in a random way and that should, if possible, be smoothed out of the fitted curve?

If the answer to the first question is yes, then the user must focus on interpolation schemes for his curve fits. If the answer to the second question is yes, then judgement is to be exercised in assessing the degree of smoothing to be applied and the choice of smoothing procedure. In this instance we refer to smoothing in the statistical sense—a sense that implies the removal of "extraneous" maxima and minima and the identification of the underlying trend with some degree of confidence. Here the user may consider the smoothing spline techniques of weighted least squares fitting described in Section 4.7 or the least squares fitting of piecewise linear $\mathscr{C}$ or piecewise cubic $\mathscr{C}^1$ functions discussed in Section 3.3.

There are also "filtering" techniques that apply to the smoothing problem. Here the data are in effect decomposed as a superposition of sinusoidal functions, and smoothing then generally consists of removing some selected high-frequency components. This is a very useful line of attack, but one that cannot be included in this work [see, for example, Gelb (1974)].

6.3 Global versus local techniques

Should the data at a given point influence the nature of the fitted curve at distant points? The answer to this is likely to depend on the user's knowledge of the physical phenomenon being modelled and on the density of data points.

If this is not thought to be an issue of significance, then it imposes no constraint on the choice of technique. If it is thought to be significant, then we should ask: How significant? If variations in the data at one point should not affect distant points, how far distant *should* the perturbation due to such a variation be propagated?

We have noted that the classical polynomial schemes and most of the spline schemes are global but that they differ in the degree of attenuation. A good intuitive picture of this characteristic is obtained by comparison of the cardinal functions for various schemes. The local schemes we have considered include the piecewise linear and cubic schemes of Chapter 3, and this is the motivation for the surface-fitting techniques to be developed in Chapter 9. For some problems the local schemes may have a decided advantage in that additional data points can be added to an interpolated curve, and the effect of this is confined to some small neighbourhood of the new points. Thus, it is not necessary to recalculate the whole of the interpolated curve.

A case in which a global technique is clearly indicated is that of analysis of "trends", which are either apparent or, for physical reasons, known to be present. For these problems, a trend in the data is usually identified, either by "eye-balling" or by analysis. The trend may be polynomial, exponential, oscillatory, etc., and when identified, a global least squares fit is made of the appropriate simple curve or surface.

6.4 Computational effort

This is an easily understood criterion whose significance will depend on the magnitude of the computation involved. In general, it is likely to become significant only for surface-fitting problems with very large numbers of data points. Precise comparisons are made more difficult by the variability of performance with different computer installations, different programming languages, different algorithms for implementation, and different levels of programming skills. However, one or two general statements can be made.

First, it is clear that, for either spline or polynomial techniques, the computational effort increases rapidly with the degree of the fitted curve. Second, fitting techniques that are local (as described in Section 6.3) in nature will generally offer significant computational advantages over global techniques.

6.5 Convergence

This is a consideration dear to the heart of the mathematician. The kinds of questions posed are as follows: Suppose that the data values correspond to a particular function f, and suppose that with n data points a curve-fitting technique determines a function g_n; with $n + 1$ points a function g_{n+1} is determined, and so on. Thus, in theory, a sequence of fitted functions $g_n, g_{n+1}, g_{n+2}, \ldots$, is determined. Can it be asserted that, in some sense, the $g_r, r = n, n + 1, \ldots$, *converge* to f? If so, how fast, and how "close" is g_r to f?

Much the greater part of the mathematical literature on curve and surface fitting is devoted to this kind of study. Its relevance for the practitioner will sometimes be unclear, especially for problems presented with a fixed number of data points. The possibility of adding more data and refining the calculations in order to take advantage of high rates of convergence may not be realistic. However, the study of convergence does give valuable insights into curve-fitting problems and can provide some further information on the choice of procedure. There is generally a trade-off to be considered between rate of convergence and volume of computation.

6.6 Visual criteria

In some fields of application, a major difficulty in designing curve- and surface-fitting techniques is the formulation of clear criteria for acceptability. For example, the intuition of the experienced geophysicist seems to involve a mixture of differentiability, smoothing, and frequency analysis all applied visually. There may also be (conscious or subconscious) considerations of bounds on the curvature and on distance from a piecewise linear interpolant to all the data points. The closer one can get to quantifying all of these visually applied criteria, the better the chances are of finding an acceptable curve- or surface-fitting method for a particular class of problems.

In this connection one may also ask which techniques lend themselves to interactive computer-aided design. There are problems for which a graphic display of a fitted curve can be utilised to apply visual criteria and then modify the curve accordingly. Certain algorithms lend themselves to such a procedure, whereas others do not. Methods based on piecewise cubic Hermite interpolation are among those that can be used in this way.

7

Surface Fitting with Polynomials

7.1 Introduction

Our treatment of curve fitting in previous chapters has not depended very heavily on the location of the data points. We have worked mostly with arbitrary data points $x_0 < \ldots < x_N$ on the line. When we consider an arbitrary set of points *in the plane*, there is not generally an obvious, unique ordering of the points, and this gives rise to some difficulties for surface fitting. The general interpolation problem in the plane might be defined as follows: Given N data points (x_i, y_i) and N numbers f_i, $i = 1, 2, \ldots, N$, find a function $f(x, y)$ from some class and defined on the whole plane (or at least a region containing the data points) for which $f(x_i, y_i) = f_i$ for $i = 1, 2, \ldots, N$.

If there is a pattern in the distribution of data points, it may be possible to use it to advantage, and in Chapter 8 we shall discuss the simplest such pattern, namely, a set of points in a rectangular lattice. In this chapter we confine our attention to introductory ideas, to constructions that will be useful in later chapters, and to the classical problem of least squares fitting of a polynomial. More sophisticated developments of least squares fitting appear in Chapter 10.

7.2 Polynomials in two variables

The role played by polynomials in surface fitting is as important as that in curve fitting, but we now have to cope with polynomials in *two* variables. We list here basis functions (monomials) in four classes of such polynomials:

Class	Basis functions
$\mathcal{P}_0$	1
$\mathcal{P}_1$	1 x y
$\mathcal{P}_2$	1 x y x^2 xy y^2
$\mathcal{P}_3$	1 x y x^2 xy y^2 x^3 x^2y xy^2 y^3

Thus, any function p in the class $\mathscr{P}_2$, for example, must be of the form

$$p(x, y) = a_1 + a_2 x + a_3 y + a_4 x^2 + a_5 xy + a_6 y^2, \qquad (7.1)$$

for some real numbers $a_1, a_2, \ldots, a_6$. More generally, $\mathscr{P}_n$ is the class of polynomials containing all functions of the form $x^i y^j$, where $0 \leqslant i + j \leqslant n$ and $i \geqslant 0, j \geqslant 0$.

Note particularly the number of basis functions in each class (this is the *dimension* of the corresponding vector space). In fact, for any n, the number of basis functions in $\mathscr{P}_n$ is $1 + 2 + 3 + \ldots + (n + 1) = \frac{1}{2}(n + 1)(n + 2)$. Once again $\mathscr{P}_0, \mathscr{P}_1, \mathscr{P}_2$, and $\mathscr{P}_3$ are known as the spaces of constant, linear, quadratic, and cubic polynomials, respectively. More specifically, the polynomials may now be described as *bivariate*.

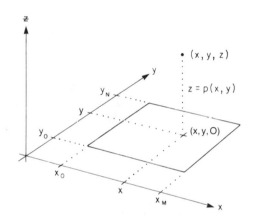

FIG. 7.1   Function defined on $[x_0, x_M] \times [y_0, y_N]$.

If we introduce a third rectangular axis (Fig. 7.1), then the equation $z = p(x, y)$ will determine a surface over the rectangle $[x_0, x_M] \times [y_0, y_N]$, a domain chosen for pictorial convenience. With each point (x, y) in the rectangle, a point with coordinates (x, y, z) on the surface is determined by putting $z = p(x, y)$. In Fig. 7.2 we illustrate two such surfaces: One corresponds to a function in $\mathscr{P}_2$, the other to one in $\mathscr{P}_3$. Note also that, if $p \in \mathscr{P}_1$, then the surface represented by $z = p(x, y) = a_1 + a_2 x + a_3 y$ is a *plane*. If $a_2 = 0$, the plane is parallel to the x-axis and if $a_3 = 0$, the plane is parallel to the y-axis.

If a vertical section is taken of a quadratic surface, the resulting curve will be a (possibly degenerate) quadratic arc. A degenerate case would be a straight line. Similarly, a vertical section of a cubic surface is generally a cubic arc, although it may "degenerate" to a quadratic arc or a straight line.

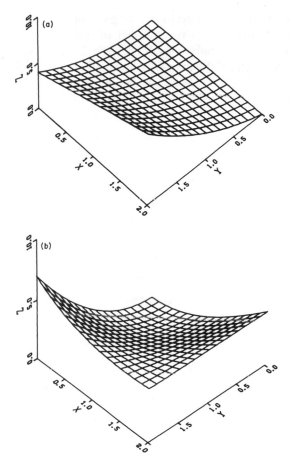

FIG. 7.2   Two polynomial surfaces: (a) $z = xy + x^2$ and (b) $z = -1 + 2x^2 + 2y^2 - x^2y - \frac{1}{3}y^3$.

7.3 Bivariate polynomial interpolation—the bad news

We first consider what is, perhaps, the most obvious surface-fitting scheme. If we have a problem with six data points, for example, we can try to fit a polynomial of class $\mathcal{P}_2$ to the data. This is a reasonable suggestion because there are six basis functions in $\mathcal{P}_2$. However, the number of data points is usually not free for choice. For example, eight data points seems to be too many for a $\mathcal{P}_2$ surface but not enough for a $\mathcal{P}_3$. One can try to get around this by taking six basis functions from $\mathcal{P}_2$ and two additional ones chosen from those in $\mathcal{P}_3$ that are not already in $\mathcal{P}_2$.

Suppose then that we succeed in deciding on an appropriate set of basis functions equal in number to the number of data points. In the case of interpolation on a line by polynomials (Section 2.7), we were able to state the theorem assuring us of the existence of a unique polynomial interpolant. *There is no such theorem for interpolation in the plane.*

To illustrate this fact consider two problems, each with four data points arranged in a square as illustrated in Fig. 7.3. For both problems we

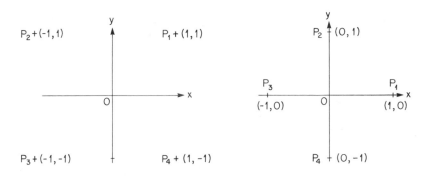

FIG. 7.3   Bilinear interpolation on four points.

make a natural choice of basis functions: $1, x, y, xy$. Let the point P_i have coordinates (x_i, y_i) for $i = 1, 2, 3, 4$, and let f_i be the assigned function value at point P_i. Now the problem is to find numbers a_1, a_2, a_3, a_4 such that

$$a_1 + a_2 x_i + a_3 y_i + a_4 x_i y_i = f_i, \qquad (7.2)$$

for $i = 1, 2, 3, 4$. If we can do so, then

$$p(x, y) = a_1 + a_2 x + a_3 y + a_4 xy \qquad (7.3)$$

will determine the interpolating surface. This function is termed *bilinear* because when either x or y is assigned a constant value, the function describes a straight line in the remaining variable. Consequently, sections of this surface by planes parallel to the xz- or yz-planes are straight lines. In the case of the first problem of Fig. 7.3, we obtain four equations for a_1, a_2, a_3 and a_4:

$$a_1 + a_2 + a_3 + a_4 = f_1,$$

$$a_1 - a_2 + a_3 - a_4 = f_2,$$

$$a_1 - a_2 - a_3 + a_4 = f_3,$$

$$a_1 + a_2 - a_3 - a_4 = f_4,$$

which have the unique solution

$$a_1 = \tfrac{1}{4}(f_1 + f_2 + f_3 + f_4), \qquad a_2 = \tfrac{1}{4}(f_1 - f_2 - f_3 - f_4),$$

$$a_3 = \tfrac{1}{4}(f_1 + f_2 - f_3 - f_4), \qquad a_4 = \tfrac{1}{4}(f_1 - f_2 + f_3 + f_4),$$

and this is the case whatever values f_1, f_2, f_3, f_4 may take. This looks encouraging. However, if we follow the same procedure in the second problem, we get into trouble. The four equations [Eq. (7.2)] are in this case,

$$a_1 + a_2 \qquad = f_1,$$

$$a_1 \qquad + a_3 = f_2,$$

$$a_1 - a_2 \qquad = f_3,$$

$$a_1 \qquad - a_3 = f_4,$$

and these are contradictory if, for example, $f_1 + f_3 \neq f_2 + f_4$. This means that, in general, for arbitrarily chosen $f_1, f_2, f_3,$ and f_4, there is *no* surface determined by a function of the form of Eq. (7.3) that will interpolate the data.

If the choice of basis functions is thought to be suspect, the next natural choice is the set $1, x, y, x^2 + y^2$. In this case it is found that, for both sets of data of Fig. 7.3, interpolants do not generally exist.

Difficulties of these kinds, coupled with the weaknesses inherited from polynomial interpolation in one dimension (cf. Section 2.2) mean that direct polynomial interpolation is not generally recommended; this is particularly so when the number of data points is large. In this case, even though an interpolating surface exists there may be very real computational problems in the form of large, ill-conditioned systems of equations to be solved. These computational problems are alleviated by the use of orthogonal polynomials [see discussion and references by I. K. Crain (1970)], but this does not overcome the more fundamentally unsatisfactory features of polynomial interpolation.

7.4 Bivariate polynomial interpolation—the good news

We have deliberately emphasized the negative aspects of polynomial interpolation in Section 7.3. It is important to recognize the limitations and possible pitfalls, but we must not throw out the baby with the bathwater. In this section we make some limited but nonetheless important positive statements that will be helpful in subsequent chapters.

Returning to the first problem of Section 7.3, we know that, at least, there is a unique bilinear interpolant $p(x, y)$ [of the form of Eq. (7.3)] to any given data at the vertices $(1, 1), (-1, 1), (1, -1), (-1, -1)$ of Problem 1 in Fig. 7.3. In fact, it is not difficult to verify that if the square is oriented as in Fig. 7.4, with centre at the origin of coordinates, then the interpolation problem using the four vertices and a bilinear interpolant has a unique solution for every θ *except* $\theta = 0$ and $\theta = \pi$. Thus, the "bad case" of Section 7.3 might be seen as *pathological* and arises from an unfortunate orientation of the square with respect to the coordinate axes.

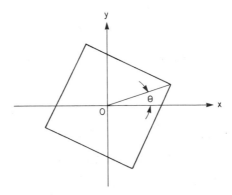

FIG. 7.4   Square with centre at the origin.

A more useful generalization of the "good case" of Section 7.3, and one which clearly avoids the angular orientation problem, is the following:

Theorem 7.4.1   *Let D denote any rectangle in the plane with sides parallel to the coordinate axes. Then there is a unique bilinear polynomial*

$$p(x, y) = a_1 + a_2 x + a_3 y + a_4 xy,$$

which interpolates prescribed data at the vertices of D.

To see this we could set up coordinates x, y in the plane of D. Let D have vertices $(x_0, y_0), (x_0, y_1), (x_1, y_1)$, and (x_1, y_0) (with $x_0 < x_1, y_0 < y_1$), and then map D onto the "standard square" with vertices $(1, 1), (-1, 1)$, $(1, -1), (1, 1)$ in the ξ, η coordinate plane (cf. Fig. 7.5) by means of the transformation

$$\xi = 1 - \frac{2(x_1 - x)}{x_1 - x_0}, \qquad \eta = 1 - \frac{2(y_1 - y)}{y_1 - y_0}. \tag{7.4}$$

Then do the interpolation on D_0, which we know to be possible, and transform the interpolating bilinear function $p_0(\xi, \eta)$ back to the x, y-coordinates by substituting from Eq. (7.4). thus, the interpolant $p(x, y)$ on

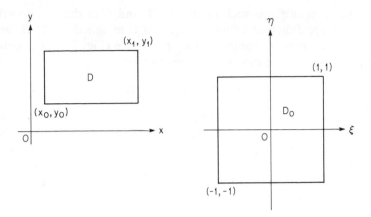

FIG. 7.5 Mapping a rectangle on the standard square.

D is given by

$$p(x, y) = p_0 \left[1 - \frac{2(x_1 - x)}{x_1 - x_0}, \quad 1 - \frac{2(y_1 - y)}{y_1 - y_0} \right]. \tag{7.5}$$

The point is that the coordinate transformations are such that the transformed function $p(x, y)$ *remains bilinear*, but in the x, y-coordinates.

This procedure may appear cumbersome, but it is a fundamental idea in the computer implementation of finite-element methods in general. We consider some of these in Chapter 9.

Let us now examine a more primitive question. Suppose we are given just three points in the plane that are not collinear. It is geometrically

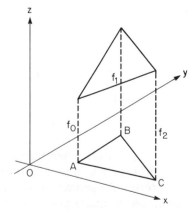

FIG. 7.6 Linear interpolation on a triangle.

obvious that, given three such points A, B, and C in the x, y-coordinate plane, the three function values f_0, f_1, and f_2 assigned to A, B, and C, respectively, determine a unique triangle in space (Fig. 7.6). The equation of the plane containing this triangle has the form

$$z = a_1 + a_2 x + a_3 y, \tag{7.6}$$

for some coefficients a_1, a_2, and a_3. If A, B, and C have coordinates (x_0, y_0), (x_1, y_1), and (x_2, y_2), respectively, then the function on the right of this equation is precisely the (bivariate) linear polynomial $p(x, y)$ for which $p(x_0, y_0) = f_0$, $p(x_1, y_1) = f_1$, and $p(x_2, y_2) = f_2$. Thus, given A, B, and C and f_0, f_1, and f_2, the coefficients a_1, a_2, and a_3 are determined by the three simultaneous equations

$$a_1 + x_0 a_2 + y_0 a_3 = f_0,$$
$$a_1 + x_1 a_2 + y_1 a_3 = f_1, \tag{7.7}$$
$$a_1 + x_2 a_2 + y_2 a_3 = f_2.$$

By Cramer's rule, these have the explicit solution in terms of determinants:

$$a_1 = \Delta^{-1} \det \begin{bmatrix} f_0 & x_0 & y_0 \\ f_1 & x_1 & y_1 \\ f_2 & x_2 & y_2 \end{bmatrix}, \qquad a_2 = \Delta^{-1} \det \begin{bmatrix} 1 & f_0 & y_0 \\ 1 & f_1 & y_1 \\ 1 & f_2 & y_2 \end{bmatrix},$$

$$a_3 = \Delta^{-1} \det \begin{bmatrix} 1 & x_0 & f_0 \\ 1 & x_1 & f_1 \\ 1 & x_2 & f_2 \end{bmatrix},$$

where

$$\Delta = \det \begin{bmatrix} 1 & x_0 & y_0 \\ 1 & x_1 & y_1 \\ 1 & x_2 & y_2 \end{bmatrix}. \tag{7.8}$$

It can be shown that $\Delta = 0$ precisely when triangle ABC has area equal to zero, and then there is no longer a unique solution for Eq. (7.7). This expresses in mathematical terms the geometrically obvious condition that, for unique linear interpolation, the data points A, B, and C must not be collinear. Some of these findings are summarized in Theorem 7.4.2.

Theorem 7.4.2 *Let T be any non-degenerate triangle in the plane. There is a unique linear polynomial*

$$p(x, y) = a_1 + a_2 x + a_3 y,$$

which interpolates prescribed data at the vertices of T.

As with bilinear interpolation on rectangles, the interpolation can always be done on a "standard" triangle and then transformed to any given triangle T. One such transformation (known as an *affine* transformation) is

$$\xi = a_{11}x + a_{12}y + b_1,$$
$$\eta = a_{21}x + a_{22}y + b_2,$$

(7.9)

where

$$a_{11} = \frac{2y_2 - y_1 - y_0}{\Delta}, \qquad a_{12} = \frac{-2x_2 + x_0 + x_1}{\Delta},$$

$$b_1 = \frac{x_2(y_1 + y_0) - y_2(x_1 + x_0)}{\Delta},$$

$$a_{21} = \frac{-\sqrt{3}(y_1 - y_0)}{\Delta}, \qquad a_{22} = \frac{\sqrt{3}(x_1 - x_0)}{\Delta},$$

$$b_2 = \frac{2}{\sqrt{3}} + \frac{\sqrt{3}(x_2(y_1 - y_0) - y_2(x_1 - x_0))}{\Delta},$$

and Δ is given by Eq. (7.8). This will map triangle T in the xy-plane (cf. Fig. 7.7) onto the standard equilateral triangle in the $\xi\eta$-plane with sides

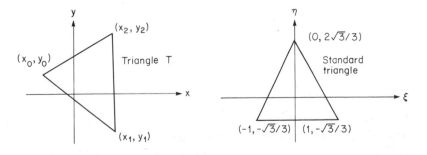

FIG. 7.7 Mapping a triangle on a standard triangle.

of length 2. The vertices are transformed as follows:

$$(x_0, y_0) \rightarrow (-1, -\sqrt{3}/3), \qquad (x_1, y_1) \rightarrow (1, -\sqrt{3}/3),$$
$$(x_2, y_2) \rightarrow (0, 2\sqrt{3}/3).$$

The mapping in the reverse direction, from the standard triangle onto T, is given by

$$x = b_{11}(\xi - b_1) + b_{12}(\eta - b_2),$$
$$y = b_{21}(\xi - b_1) + b_{22}(\eta - b_2),$$
(7.10)

where

$$b_{11} = \tfrac{1}{2}(x_1 - x_0), \qquad b_{12} = \frac{1}{2\sqrt{3}}(2x_2 - x_0 - x_1),$$

$$b_{21} = \tfrac{1}{2}(y_1 - y_0), \qquad b_{22} = \frac{1}{2\sqrt{3}}(2y_2 - y_0 - y_1).$$

and b_1, b_2 are as in Eq. (7.9).

EXAMPLE   Let T be the triangle with vertices $(-1, 2)$, $(-2, 3)$, and $(1, 4)$ in the x, y-coordinate plane. It is found that $\Delta = -4$, and Eq. (7.9) becomes

$$\xi = -\frac{3}{4}x + \frac{5}{4}y - \frac{17}{4},$$

$$\eta = \frac{\sqrt{3}}{4}x + \frac{\sqrt{3}}{4}y - \frac{7\sqrt{3}}{12}.$$

It is easily verified that $(-1, 2) \rightarrow (-1, -\sqrt{3}/3)$, $(-2, 3) \rightarrow (1, -\sqrt{3}/3)$, and $(1, 4) \rightarrow (0, 2\sqrt{3}/3)$.

The reverse, or inverse transformation, is

$$x = -\frac{1}{2}\xi + \frac{5\sqrt{3}}{6}\eta - \frac{2}{3},$$

$$y = \frac{1}{2}\xi + \frac{\sqrt{3}}{2}\eta + 3,$$

and maps the standard triangle onto T.

If function values f_0, f_1, f_2 are assigned to the vertices of a general triangle T, then to find the value of the linear interpolant at a point (x_0, y_0) of T, Eq. (7.9) is used to find the corresponding point (ξ_0, η_0) of the standard triangle. The linear interpolant $p_0(\xi, \eta)$ is found on *this* triangle and evaluated at (ξ_0, η_0). The interpolant on T then takes this same value, $p_0(\xi_0, \eta_0)$ at (x_0, y_0). Thus, [cf. Eq. (7.5)],

$$p(x, y) = p_0(a_{11}x + a_{12}y + b_1, \quad a_{21}x + a_{22}y + b_2),$$

where the a and b coefficients are as in Eq. (7.9). It is easily verified that the linear interpolant $p_0(\xi, \eta)$ on the standard triangle determines a linear interpolant $p(x, y)$ on T.

As in the case of rectangles, this procedure of transforming to a standard set of vertices comes into its own when interpolation on many different triangles is required. It also increases the efficiency of computation more dramatically in the case of interpolation with polynomials of higher degree, and that is the topic to be introduced next in Sections 7.5 and 7.6.

7.5 Higher degree interpolation on triangles

Consider first the possibility of interpolating on any triangle T with polynomials containing some of the monomials x^2, xy, y^2 from $\mathcal{P}_2$. If the process is to be valid under affine transformations mapping an arbitrary triangle onto a standard triangle, it is easily seen that all of these three monomials must be admitted. That is, we must look for interpolation with (possibly) "complete" quadratic functions from $\mathcal{P}_2$. Since $\mathcal{P}_2$ has dimension six (recall that it is spanned by the six monomials 1, x, y, x^2, xy, and y^2), we must seek a configuration of six nodes on a triangle, i.e. points at which data are to be assigned. Symmetry considerations then suggest that the vertices of T and points in the middle of each side be used as nodes. It turns out that *there is, indeed, a unique polynomial in $\mathcal{P}_2$ taking assigned values at the vertices and mid-sides of a triangle T.*

For reasons that will be made clear in Chapter 10, this result is not very useful in large-scale surface fitting (i.e. with *many* data points). A similar result applies for polynomials from $\mathcal{P}_3$ and the set of nodes of interpolation indicated in Fig. 7.8. The sides are trisected by nodes of interpolation, and the centroid of T determines the tenth node required to match the dimensionality of $\mathcal{P}_3$. Like the so-called "quadratic triangle", this will not be very useful for us, but let us take advantage of this example to illustrate the possibility of using derivative nodes as in the case of Hermite interpolation on the line.

It turns out that the precise location of the six nodes in Fig. 7.8a on the edges (and not at the vertices) is not important. They can be moved back and forth along the edges without upsetting the fundamental statement on the existence of a unique interpolant from $\mathcal{P}_3$. Now suppose that on each side of the triangle one interior node approaches each vertex. Thus, three nodes converge onto each vertex of T. It is plausible that, in the limit, we obtain one function-value node at each vertex and *two (partial) derivative nodes* at each vertex. (See Section 8.4 for a formal definition of the term "node".) In fact, what is sometimes known as the "Hermite form of the cubic triangle" is arrived at in this way.

If we are interpolating to a function $f(x, y)$ on T, the values of f to be reproduced by a polynomial $p \in \mathcal{P}_3$ are the values of f and of the first partial derivatives f_x, f_y at each vertex and the value of f at the centroid of T. This nodal configuration is indicated in Fig. 7.8b, where the circled dot at each vertex means *three* nodal values are to be used, namely, the values of f, f_x, and f_y. Once again, there is a unique polynomial in $\mathcal{P}_3$ taking assigned values at these ten nodes of interpolation.

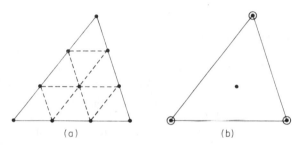

FIG. 7.8   Interpolation on a triangle with functions from P_3.

These constructions can be continued to admit interpolation with higher degree polynomials at the expense of introducing more points of interpolation or higher order derivatives of interpolation. As a final example of this kind, note the existence of a "complete quintic triangle". There are 21 nodes defined by values of $f, f_x, f_y, f_{xx}, f_{xy}$ and f_{yy} at each of three vertices together with *normal* derivatives $\partial f/\partial n$ at the mid-point of each side. These match the 21 monomials $x^j y^k$, where $0 \leqslant j + k \leqslant 5$ in $\mathcal{P}_5$. The nodal configuration is indicated schematically in Fig. 7.9. The position of this interpolation scheme in a family of interpolation schemes of high polynomial degree is discussed by Morgan and Scott (1975). A modified scheme consisting of 18 nodes and reproducing functions in $\mathcal{P}_4$ has been devised by Mitchell (1973).

For the purposes of surface construction there is good reason to avoid the use of derivative nodes of order two or more if possible and hence the complete quintic triangle. Another criterion, depending on the nature of interpolants on *adjacent* triangles, appears in Chapter 9 and generally disqualifies the other schemes developed in this section. More suitable and more complicated procedures for interpolation on triangles (as a tool for surface construction) are presented in that chapter.

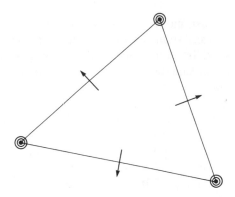

FIG. 7.9   Nodal configuration for complete quintic triangle.

7.6 Higher degree interpolation on rectangles

In Sections 7.4 and 7.5, interpolation schemes on triangles using complete polynomials in $\mathscr{P}_1$, $\mathscr{P}_2$, $\mathscr{P}_3$, or $\mathscr{P}_5$ have been discussed. These have the useful property that the interpolating polynomial function values are invariant under arbitrary affine transformations [as in Eq. (7.9)] of the underlying triangle. Thus, the triangle can be translated and rotated in the plane and stretched or contracted in the directions of the axes without disturbing the value of the interpolating polynomial at corresponding points. As the discussion of Sections 7.3 and 7.4 might lead us to expect, interpolation on rectangles is not well matched with the use of *complete* polynomials. It seems that the bilateral symmetries of the rectangle do not lend themselves to distributions of nodes in the numbers $3, 6, 10, 15,$ $21, \ldots, \frac{1}{2}n(n + 1), \ldots$ These are, of course, the dimensions of $\mathscr{P}_1$, $\mathscr{P}_2$, $\mathscr{P}_3, \ldots$

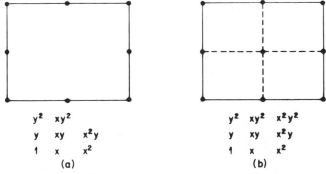

FIG. 7.10   Two quadratic interpolation schemes.

So, in general, the best that we can hope for is invariance of interpolants on rectangles under translation parallel to the axes and changes of scale. These are represented by equations of the form $\xi = a_1 x + b_1$ and $\eta = a_2 y + b_2$, with $a_1 a_2 \neq 0$. Equations (7.4), which map an arbitrary rectangle onto a standard rectangle, are of this type.

One of the most primitive interpolation schemes on rectangles is, of course, that using bilinear polynomials and described in Theorem 7.4.1. There are two interpolation schemes on rectangles that are precise for all polynomials in $\mathscr{P}_2$ (and not $\mathscr{P}_3$) and which must, therefore, have at least six nodes. They are indicated schematically, together with a basis of monomial functions, in Fig. 7.10. That of Fig. 7.10b is known as the "biquadratic rectangle".

For interpolation with cubic polynomials, it turns out that at least 12 nodes are required (the dimension of $\mathscr{P}_3$ is ten). At this point, the possibility of interpolation of Hermite type arises. The 12 nodes indicated in Fig. 7.11a are a natural choice and they do, indeed, match the 12 monomials in the array

$$
\begin{array}{llll}
y^3 & xy^3 & & \\
y^2 & xy^2 & & \\
y & xy & x^2 y & x^3 y \\
1 & x & x^2 & x^3
\end{array}
$$

As in the discussion of the complete cubic in Section 7.5, it turns out that the locations of the nodes on the (interiors of the) sides of the rectangle in Fig. 7.11a are not critical. Once this is recognized it is not surprising that the nodes indicated in Fig. 7.11b also form a viable interpolation scheme together with the same monomials of the above array (see also Section 9.2). Recall that, in this illustration, a circled dot at a point implies that *three* nodes, the values of f, f_x, and f_y at this point, are to be used.

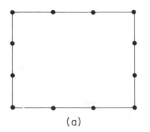
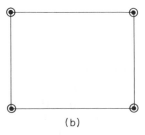

(a)   (b)

FIG. 7.11   Two cubic interpolation schemes.

The place of these interpolation methods in the context of families of interpolating schemes of higher degree on rectangles has been developed by Melkes (1972) and Lancaster and Watkins (1977).

A final, and particularly interesting example, is indicated in Fig. 7.12. There are 16 nodes consisting of the values of f, f_x, f_y, and f_{xy} at each vertex. The array of 16 monomials associated with these nodes is also indicated in

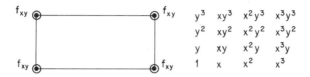

$$
\begin{array}{cccc}
y^3 & xy^3 & x^2y^3 & x^3y^3 \\
y^2 & xy^2 & x^2y^2 & x^3y^2 \\
y & xy & x^2y & x^3y \\
1 & x & x^2 & x^3
\end{array}
$$

FIG. 7.12 The bicubic rectangle.

Fig. 7.12. Observe that their span includes all of $\mathscr{P}_3$. Any polynomial obtained as a linear combination of these monomials is called a *bicubic* polynomial and, consequently, this combination of polynomials and nodes of interpolation is known as the *bicubic rectangle*. Further discussion of the bicubic rectangle appears in Chapter 8, and it will be found useful in the constructions of Chapter 9.

The construction of interpolation schemes on triangles, rectangles, and other figures in two or more dimensions plays an important role in the finite-element method for the numerical solution of partial differential equations [see, for example, Strang and Fix (1973) and Ciarlet (1978)]. However, in that context, inter-element smoothness is not as important as it is for us. This issue will be one of our main concerns in Chapter 9.

7.7 Least squares fitting of polynomial surfaces

In Sections 2.5–2.8, curve fitting by the least squares method was discussed. The intuitive ideas in the context of surface fitting are the same. Thus, as pointed out in Section 2.8, least squares fits generally involve either *smoothing* of the data or identifying an apparent *trend* in the data. However, as in the case of fitting a polynomial surface by interpolation, there are certain inherent difficulties in fitting a polynomial surface by the least squares methods. For example, after reading Section 7.7, the reader may wish to try fitting a surface determined by a linear combination of functions $1, x, y, xy$ to data at the five points $(0, 0), (1, 0), (0, 1), (-1, 0), (0, -1)$. From the computational point of view, it is not these pathological cases that give trouble, but the cases where the data points lie *close* to a pathological example, in which case ill-conditioned problems may result. However, we disregard these difficulties for the time being and proceed with a formal analysis.

We assume that values f_i are given at $N + 1$ arbitrary data points (x_i, y_i), $i = 0, 1, \ldots, N$, and for the purpose of illustration, we suppose that $N \geqslant 5$, and we fit a quadratic surface to the data. Then, the surface is to be determined by a function having the form

$$p(x, y) = a_1 + a_2 x + a_3 y + a_4 x^2 + a_5 xy + a_6 y^2, \tag{7.11}$$

for some choice of $a_1, a_2, \ldots, a_6$. At each data point the difference between the surface elevation and f_i is $p(x_i, y_i) - f_i$. We are then to adjust p (by choice of $a_1, \ldots, a_6$) so as to minimize

$$E(p) = \sum_{i=0}^{N} (p(x_i, y_i) - f_i)^2, \tag{7.12}$$

which should be compared with Eq. (2.10). The technique is then just that used in Section 2.5. The function E of $a_1, \ldots, a_6$ will have a minimum only when $\partial E/\partial a_i = 0$ for $i = 1, 2, \ldots, 6$. These conditions yield six linear equations in the unknowns $a_1, \ldots, a_6$ (the normal equations). We omit the details and assert that the equations turn out to be

$$(N + 1)a_1 + (\Sigma x_i)a_2 + (\Sigma y_i)a_3 + (\Sigma x_i^2)a_4$$
$$+ (\Sigma x_i y_i)a_5 + (\Sigma y_i^2)a_6 = \Sigma f_i,$$

$$(\Sigma x_i)a_1 + (\Sigma x_i^2)a_2 + (\Sigma x_i y_i)a_3 + (\Sigma x_i^3)a_4$$
$$+ (\Sigma x_i^2 y_i)a_5 + (\Sigma x_i y_i^2)a_6 = \Sigma x_i f_i,$$

$$(\Sigma y_i)a_1 + (\Sigma x_i y_i)a_2 + (\Sigma y_i^2)a_3 + (\Sigma x_i^2 y_i)a_4$$
$$+ (\Sigma x_i y_i^2)a_5 + (\Sigma y_i^3)a_6 = \Sigma y_i f_i, \tag{7.13}$$

$$(\Sigma x_i^2)a_1 + (\Sigma x_i^3)a_2 + (\Sigma x_i^2 y_i)a_3 + (\Sigma x_i^4)a_4$$
$$+ (\Sigma x_i^3 y_i)a_5 + (\Sigma x_i^2 y_i^2)a_6 = \Sigma x_i^2 f_i,$$

$$(\Sigma x_i y_i)a_1 + (\Sigma x_i^2 y_i)a_2 + (\Sigma x_i y_i^2)a_3 + (\Sigma x_i^3 y_i)a_4$$
$$+ (\Sigma x_i^2 y_i^2)a_5 + (\Sigma x_i y_i^3)a_6 = \Sigma x_i y_i f_i,$$

$$(\Sigma y_i^2)a_1 + (\Sigma x_i y_i^2)a_2 + (\Sigma y_i^3)a_3 + (\Sigma x_i^2 y_i^2)a_4$$
$$+ (\Sigma x_i y_i^3)a_5 + (\Sigma y_i^4)a_6 = \Sigma y_i^2 f_i,$$

and all the summations are for i from zero to N.

To obtain the normal equations for fitting a polynomial of first degree (a plane surface), simply take the first three equations and put $a_4 = a_5 = a_6 = 0$. Similarly, for a fitted bilinear surface, take the first, second, third, and fifth equations and set $a_4 = a_6 = 0$.

As in the case of one variable, the normal equations are conveniently expressed in matrix–vector form. In the case of the full set of Eq. (7.13), we introduce the $(N + 1) \times 6$ matrix

$$V = \begin{bmatrix} 1 & x_0 & y_0 & x_0^2 & x_0 y_0 & y_0^2 \\ 1 & x_1 & y_1 & x_1^2 & x_1 y_1 & y_1^2 \\ \vdots & & & & & \vdots \\ 1 & x_N & y_N & x_N^2 & x_N y_N & y_N^2 \end{bmatrix},$$

and the column vectors,

$$\mathbf{a} = \begin{bmatrix} a_1 \\ a_2 \\ a_3 \\ a_4 \\ a_5 \\ a_6 \end{bmatrix}, \quad \mathbf{f} = \begin{bmatrix} f_0 \\ f_1 \\ \vdots \\ f_N \end{bmatrix}.$$

Then it is easily verified that Eq. (7.13) takes the form

$$V^T V \mathbf{a} = V^T \mathbf{f}. \tag{7.14}$$

Comparison should be made with Eq. (2.14) for the one-variable case.

Solution of Eq. (7.14) or Eq. (7.13) for $\mathbf{a}$ now determines the polynomial of Eq. (7.11) that best fits the data in the sense of least squared deviations. The cautionary remarks made at the end of Section 2.6 with regard to large-scale computation apply equally well here.

Note that if there are only six points of interpolation, then $N + 1 = 6$, and so V is square. If the six points are on the mid-sides and vertices of a triangle, then (cf. Section 7.5) V and $V^T V$ will be *nonsingular*. That is, $(V^T V)^{-1}$ exists, Eq. (7.14) has the unique solution $\mathbf{a} = (V^T V)^{-1} V^T \mathbf{f}$, and the corresponding function $p(x, y)$ is the *interpolant* to the data; the deviations are now zero.

More generally, if the $N + 1$ points $(x_0, y_0), \ldots, (x_N, y_N)$ $(N \geqslant 5)$ contain a subset of six points on which a unique quadratic interpolant exists, then $(V^T V)^{-1}$ exists (V has "full rank"), and the least squares problem has a unique solution.

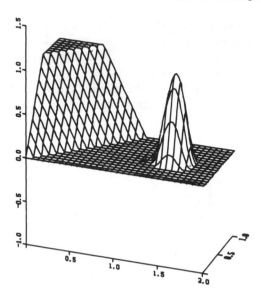

FIG. 7.13   Model problem for surface fitting.

For illustrating surface-fitting techniques, we shall often use a standard example. The model used is illustrated in Fig. 7.13 and represents a mountain on a plane and a ramp leading to another plane. Clearly, the two examples used in Chapters 2 and 4 are sections of our surface. The surface specifies the underlying function $f(x, y)$, which we sample at various sets of data points and then attempt to represent by various surface-fitting techniques.

The surface heights z of the model problem are defined more precisely by

$$
\begin{aligned}
&z = 1, &&\text{if } \; y - x \geqslant \tfrac{1}{2}, \\
&z = 2(y - x), &&\text{if } \; 0 \leqslant y - x \leqslant \tfrac{1}{2}, \\
&z = \tfrac{1}{2}\{\cos(4\pi[(x - \tfrac{3}{2})^2 && \\
&\quad + (y - \tfrac{1}{2})^2]^{1/2}) + 1\}, &&\text{if } \; [(x - \tfrac{3}{2})^2 + (y - \tfrac{1}{2})^2] \leqslant \tfrac{1}{16}, \\
&z = 0, &&\text{otherwise,}
\end{aligned}
\tag{7.15}
$$

and the boundary of the region is the rectangle with vertices $(0, 0)$, $(2, 0)$, $(2, 1)$, $(0, 1)$. This model is deliberately chosen to have two "creases" (where discontinuities in the first derivatives occur). It is interesting to see how different interpolation schemes cope with such features. In particular, schemes for $\mathscr{C}^1$ interpolating surfaces can be expected to run into difficulties here.

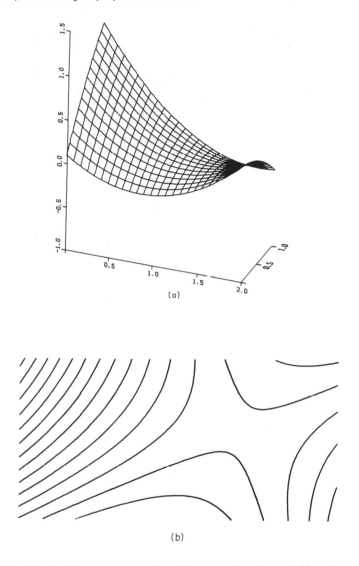

FIG. 7.14   Quadratic least squares surface: (a) perspective view and (b) contour map.

Figure 7.14 shows the quadratic surface determined by solving Eq. (7.13) using $N = 150$ randomly spaced data points. The function values at these points are determined by the model problem of Fig. 7.13. Note that the quadratic surface is unable to "follow" the mountain, and the net effect is simply to smooth out this feature.

8

Surface Interpolation by Tensor Product and Blending Methods

8.1 Introduction

In this chapter the first techniques for surface construction with large data sets are presented. It is assumed that data points lie on a complete rectangular lattice in the xy-coordinate plane and function values (and possibly slopes) are assigned to each point. Thus, if there are $(M + 1) \times (N + 1)$ points in the lattice, $M + 1$ in one direction and $N + 1$ in the other, the coordinate axes are supposed to be drawn parallel to the edges of the lattice. Then each point is assigned coordinates (x_i, y_j), where i takes values $0, 1, 2, \ldots, M$, j takes values $0, 1, \ldots, N$, and $x_0 < x_1 < \ldots < x_M$ and $y_0 < y_1 < \ldots < y_N$. Given such an array of data points and numbers f_{ij}, $0 \leq i \leq M$ and $0 \leq j \leq N$, the interpolation problem would require the determination of a function f of x and y defined on the rectangle $[x_0, x_M] \times [y_0, y_N]$, for which $f(x_i, y_j) = f_{ij}$ for each i and j.

We shall describe a rectangular lattice as *uniform* if the x_i's are uniformly spaced and the y_j's are uniformly spaced.

It is clear that, in many practical problems, the data points are not so conveniently arranged. However, it is important that we treat this case first before going on to the more difficult problems with less regularity in the data points. Also, it is often the case in practice that data points are *close* to some regular pattern. We may then wish to do some preliminary processing to produce modified data which does lie in a manageable and useful configuration. Two such techniques are discussed in Chapter 11.

8.2 Product schemes

The basic idea of this chapter is to take advantage of interpolation methods already discussed with data given *on a line* and, by first considering interpolation along the lines of a rectangular lattice parallel to the coordinate axes, to combine them somehow into a function defined over the whole rectangle $[x_0, x_M] \times [y_0, y_N]$ of the introduction.

We shall consider first the relatively simple case in which data is given only in the form of function values at the grid points. In Section 8.4 we deal with more general product interpolation schemes which admit derivative data as well.

First, it is assumed that we have available a scheme for *curve* fitting by interpolation. Several candidates for this have been developed in Chapters 2–4. The curve-fitting technique may be that of polynomial interpolation or natural cubic spline interpolation, for example. Whichever technique we have in mind, we suppose that a set of *cardinal basis* functions has been constructed. Thus, if the data points on the line are $x_0 < x_1 < \ldots < x_M$, then the *i*th cardinal basis function is defined by setting $f_i = 1$ and $f_j = 0$ if $j \neq i$ for $j = 0, 1, \ldots, M$. In this way, $M + 1$ basis functions are generated, which we call $\varphi_0, \varphi_1, \ldots, \varphi_M$. Then use the same technique to construct cardinal basis functions $\psi_0, \psi_1, \ldots, \psi_N$ on the data points $y_0, y_1, \ldots, y_N$.

Now we form products of these basis functions. Consider the set of $(M + 1)(N + 1)$ functions of the form

$$c_{ij}(x, y) = \varphi_i(x)\psi_j(y), \quad i = 0, 1, \ldots, M, \quad j = 0, 1, \ldots, N. \quad (8.1)$$

It is clear from the definition of the φ_i and ψ_i that these functions satisfy the following conditions at the points of the lattice:

$$c_{ij}(x_k, y_l) = \varphi_i(x_k)\psi_j(y_l) = \begin{cases} 1 & \text{if } k = i \text{ and } l = j, \\ 0 & \text{otherwise.} \end{cases} \quad (8.2)$$

So the functions c_{ij} also behave like cardinal functions but are based now on the rectangular lattice. Each c_{ij} takes the value 1 at one and only one point of the lattice, the point (x_i, y_j), and is zero at all of the other points.

As an illustration, we suppose that the curve-fitting technique chosen is natural cubic spline interpolation, as described in Section 4.2. Some cardinal basis functions (φ_i or ψ_j) on a set of seven points are illustrated in Fig. 4.2. Two product cardinal functions on a rectangular grid of $5 \times 9 = 45$ points are illustrated in Fig. 8.1.

A function that takes the value f_{ij} at (x_i, y_j) and the value zero at all other data points is then the function $f_{ij}c_{ij}(x, y)$, and so a function taking the value f_{ij} at (x_i, y_j) for every i and j is

$$P(x, y) = \sum_{j=0}^{N} \sum_{i=0}^{M} f_{ij}c_{ij}(x, y). \quad (8.3)$$

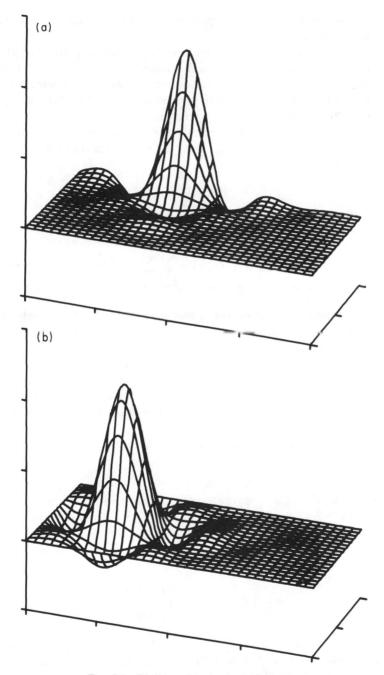

FIG. 8.1 Bicubic spline cardinal functions.

Thus, we have constructed an interpolant to the data, and $z = P(x, y)$ will determine the fitted surface.

Before going on to provide illustrations, one further point should be made. In this discussion we have used *cardinal* basis functions. It is not necessary to use cardinal functions to develop the product scheme; any set of basis functions for the curve-fitting process could be used. For example, we described in Section 4.3 the construction of *B*-spline basis functions. Utilizing this idea, we might construct bases of *B*-splines $\hat{\varphi}_0, \ldots, \hat{\varphi}_M$ on $x_0, \ldots, x_M$ and $\hat{\psi}_0, \ldots, \hat{\psi}_N$ on $y_0, \ldots, y_N$. Product functions would then be of the form

$$\hat{c}_{ij}(x, y) = \hat{\varphi}_i(x)\hat{\psi}_j(y),$$

but would not satisfy the cardinality functions of Eq. (8.2).

Nevertheless, we consider a surface defined by a function of the form

$$P(x, y) = \sum_{j=0}^{N} \sum_{i=0}^{M} \alpha_{ij}\hat{c}_{ij}(x, y) = \sum_{j=0}^{N} \sum_{i=0}^{M} \alpha_{ij}\hat{\varphi}_i(x)\hat{\psi}_j(y), \qquad (8.4)$$

for some $(M + 1)(N + 1)$ real numbers α_{ij}. It can be proved that the α_{ij}'s can be determined uniquely in such a way that $P(x_i, y_i) = f_{ij}$ for each i and j. The surface fitted would be exactly that of Eq. (8.3)—it would only be described in a different way.

A surface defined in this way [using Eq. (8.3) or Eq. (8.4), for example] from natural cubic splines will be called a *bicubic spline surface* (although the term is used elsewhere for splines which do not necessarily satisfy the "natural" boundary conditions), and it will represent a $\mathscr{C}^2$ function on the defining rectangle.

We state a general theorem containing the essentials of all product schemes.

Theorem 8.2.1   *Let $\varphi_0, \ldots, \varphi_M$ be a set of functions and $x_0 < x_1 < \ldots < x_M$ a set of points with the property that, for any $f_0, \ldots, f_M$, there exist unique numbers $\alpha_0, \alpha_1, \ldots, \alpha_M$ such that $\sum_{i=0}^{M} \alpha_i\varphi_i(x_j) = f_j$ for $j = 0, 1, \ldots, M$. Let $\psi_0, \psi_1, \ldots, \psi_N$ have the corresponding property with respect to points $y_0 < y_1 < \ldots < y_N$ and define*

$$c_{ij}(x, y) = \varphi_i(x)\psi_j(y), \qquad \begin{cases} i = 0, 1, \ldots, M, \\ j = 0, 1, \ldots, N. \end{cases}$$

Then, given any set of numbers f_{ij} there exists a unique corresponding set of numbers α_{ij} such that the function $P(x, y) = \sum_j \sum_i \alpha_{ij}c_{ij}(x, y)$ satisfies the interpolatory conditions $P(x_i, y_j) = f_{ij}$ for each i and j.

We chose to introduce the product functions via the cardinal bases because this seems the most intuitive and easily visualized approach. However, cardinal bases are not necessarily the best for computational purposes. The evaluation and storage of the cardinal functions φ_i and ψ_j themselves can be very costly. For spline techniques with large numbers of data points, the use of B-spline bases is likely to be most reliable from the computational point of view. [For more details on the use and application of bicubic spline surfaces, see, for example, de Boor (1962, 1978), Bhattacharyya (1969), and Späth (1969, 1974).]

8.3 Illustrations of product schemes

EXAMPLE 1   The simplest possible rectangular lattice has precisely two points in each direction. Thus, there are four points at the vertices of a rectangle with sides parallel to the coordinate axes (Fig. 8.2). For interpolation in the x- and y-directions, use linear polynomial interpolation. Thus, the cardinal functions are

$$\varphi_0(x) = (x_1 - x)/(x_1 - x_0), \qquad \varphi_1(x) = (x - x_0)/(x_1 - x_0), \quad (8.5)$$
$$\psi_0(y) = (y_1 - y)/(y_1 - y_0), \qquad \psi_1(y) = (y - y_0)/(y_1 - y_0). \quad (8.6)$$

There are four basis functions for interpolation in the plane at these four points given by Eq. (8.1):

$$c_{00}(x, y) = \varphi_0(x)\psi_0(y) = \left(\frac{x_1 - x}{x_1 - x_0}\right)\left(\frac{y_1 - y}{y_1 - y_0}\right),$$

$$c_{01}(x, y) = \varphi_0(x)\psi_1(y) = \left(\frac{x_1 - x}{x_1 - x_0}\right)\left(\frac{y - y_0}{y_1 - y_0}\right),$$

$$c_{10}(x, y) = \varphi_1(x)\psi_0(y) = \left(\frac{x - x_0}{x_1 - x_0}\right)\left(\frac{y_1 - y}{y_1 - y_0}\right),$$

$$c_{11}(x, y) = \varphi_1(x)\psi_1(y) = \left(\frac{x - x_0}{x_1 - x_0}\right)\left(\frac{y - y_0}{y_1 - y_0}\right).$$

Note that all of these cardinal functions are bilinear, i.e. they have the form $a_1 + a_2 x + a_3 y + a_4 xy$ for some constants a_1, a_2, a_3, a_4 (depending on i and j). Thus, this product scheme produces a basis of cardinal functions for bilinear interpolation on the rectangle of Theorem 7.4.1.

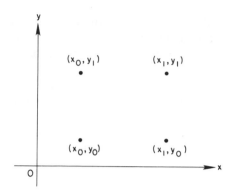

FIG. 8.2   Primitive lattice.

EXAMPLE 2   Suppose now that the lattice is 3 × 3, and we form a product scheme using quadratic Lagrangian interpolation in each direction. Thus, the cardinal basis functions $\varphi_0(x)$, $\varphi_1(x)$, $\varphi_2(x)$ and $\psi_0(y)$, $\psi_1(y)$, $\psi_2(y)$ are quadratic polynomials. Their formulation (on a special set of abscissas) has been considered in detail in Section 2.3.

Then the nine basis functions $c_{ij}(x, y)$ of Eq. (8.1) are *biquadratic* polynomials, i.e. they are linear combinations of monomials $x^i y^j$ with $1 \leq i + j \leq 2$. If the points happen to form a uniform lattice, then this construction provides a cardinal basis for the "biquadratic rectangle" of Fig. 7.10b.

EXAMPLE 3   Suppose that the lattice has $M + 1$ points $x_0, x_1, \ldots, x_M$ in the x-direction and just two, y_0, y_1, in the y-direction. Then there are just two cardinal basis functions in the y-direction $\psi_0(y)$ and $\psi_1(y)$ and, combining Eqs. (8.2) and (8.3), the interpolating function has the form

$$P(x, y) = \sum_{i=0}^{M} f_{i0}\varphi_i(x)\psi_0(y) + \sum_{i=0}^{M} f_{i1}\varphi_i(x)\psi_1(y),$$

where $\varphi_0(x), \ldots, \varphi_M(x)$ are the cardinal basis functions in the x-direction.

Linear interpolation in the y-direction would imply the use of the functions ψ_0, ψ_1 of Eq. (8.6). In this case, the surface can be visualized by first considering the x-interpolants on the lines $y = y_0$, $y = y_1$ (cf. Fig. 8.3, where $M = 3$). These two curves are then joined by straight-line segments in the y-direction, thus generating a "ruled" surface. This procedure is sometimes known as "railing" between the two curves on $y = y_0$ and $y = y_1$.

If piecewise linear interpolation is used in the x-direction, then the result on each rectangle $[x_{i-1}, x_i] \times [y_0, y_1]$ is simply a bilinear interpolant again, as in the first example above. These bilinear patches are joined on the lines $x = x_1$, $x = x_2$, ..., $x = x_{M-1}$ in such a way as to generate a continuous surface; in fact, a $\mathscr{C}^0$ surface, as defined in Section 1.6.

More generally, the smoothness of such a ruled surface is limited only by that of the interpolation scheme in the x-direction. If Lagrangian polynomial or cubic spline procedures are used in the x-direction, for example, then the surfaces generated will generally be of classes $\mathscr{C}^\infty$ and $\mathscr{C}^2$, respectively (cf. Section 1.6).

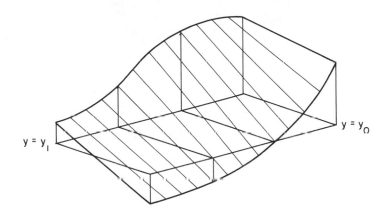

FIG. 8.3   Tensor product interpolant—linear in the y direction.

EXAMPLE 4   Now consider the model surface of Fig. 7.13. The procedure is to form a uniform lattice on the rectangular base, sample function values at the lattice points from the heights of the model surface, and construct an interpolating surface. Comparisons are made among the results of some such interpolation procedures.

Lagrangian polynomial interpolation is used first on lattices of $5 \times 3 = 15$ points and $9 \times 5 = 45$ points. The results are illustrated in Fig. 8.4. The lower degree surface indicates the main features of the model surface reasonably well. In the second case, the main features are still visible, but the increased degree of the polynomial produces large anomalies near the edges. These polynomial surfaces are in marked contrast to the following results using the same numbers of data points but cubic spline interpolation in place of Lagrangian polynomial interpolation.

Six bicubic spline interpolating surfaces are shown in Figs. 8.5 and 8.6. The first three are on uniform lattices with $3 \times 5 = 15$, $5 \times 9 = 45$, $9 \times 17 = 153$ points, respectively. The second three are on rectangular lattices with corresponding numbers of points, but in this case, the x_i's are chosen

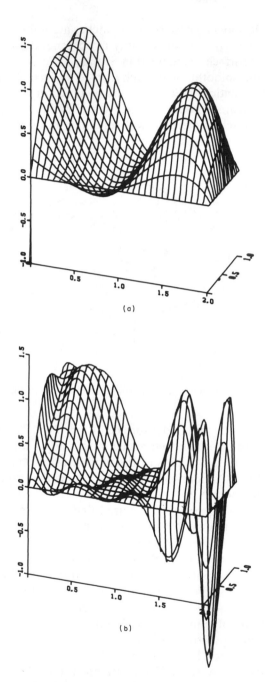

FIG. 8.4   Two product polynomial interpolants: (a) 15 points and (b) 45 points.

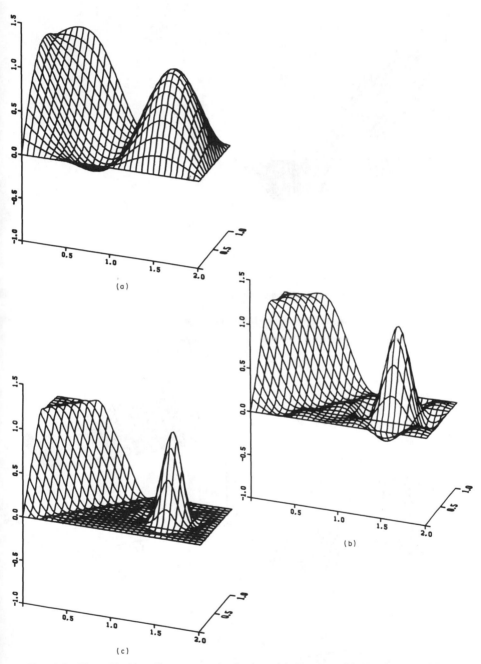

FIG. 8.5 Three bicubic splines on regular lattices: (a) 15 points, (b) 45 points and (c) 153 points.

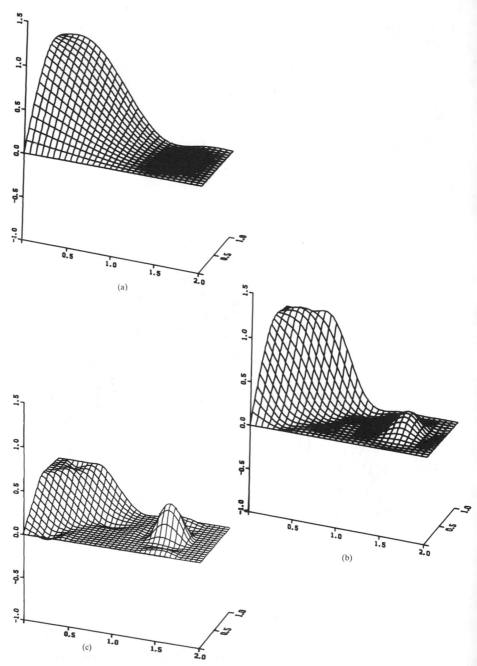

Fig. 8.6 Bicubic spline interpolating surfaces: (a) 15 points, (b) 45 points and (c) 153 points.

randomly, as are the y_j's. In Fig. 8.6a the points of interpolation chosen "missed" the mountain almost completely so that this feature is hardly visible.

In Fig. 8.7 we illustrate a contour diagram of the bicubic surface in Fig. 8.6c. This illustrates very clearly that anomalies may arise which are peculiar to the technique involved and bear no relation to the data. In particular, this makes clear a "dimpling" phenomenon, which is based on the underlying grid.

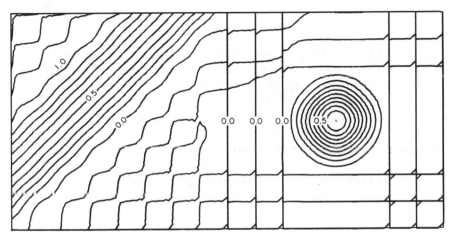

FIG. 8.7   Contour map of bicubic surface of Fig. 8.6c.

8.4 Product schemes with derivative data

This section is devoted to the problem of interpolation on a rectangular lattice in the case when we are given not only function values but derivative values as well. In particular, the definitions and theorems given will allow us to extend the piecewise Hermitian interpolation procedure of Section 3.7 to surface fitting. In order to do this, it is useful to make a more precise definition of the term *node*, first encountered in Chapter 3. The first step is to define a node in the context of *curve* fitting.

Definition   *If a function f is known to belong to the class $\mathscr{C}^k[a, b]$, and if the value of $f^{(i)}(u)$ is given for some i, $0 \le i \le k$, and $u \in [a, b]$, then the ordered pair $(u; i)$ is said to be a node. We refer to $f^{(i)}(u)$ as the nodal value of f at this node.*

These conventions are illustrated in the two following descriptions of interpolation processes on an interval.

EXAMPLE 1   *Simple Interpolation.*   Here we are looking for a continuous function taking on prescribed values at the $N + 1$ distinct points $x_0, x_1, \ldots, x_N$. Therefore the space $\mathscr{C}[a, b]$ is appropriate, and the nodes are $(x_j; 0)$, $j = 0, 1, \ldots, N$, indicating that zeroth-order derivatives (function values) are given at $x_0, x_1, \ldots, x_N$.

EXAMPLE 2   *Simple Hermitian Interpolation.*   Now, slopes as well as function values are given at the distinct points $x_0, x_1, \ldots, x_N$. The interpolating function must be differentiable, so we work with functions from $\mathscr{C}^1[a, b]$. The nodes are $(x_j; 0)$, $j = 0, 1, \ldots, N$, and $(x_j; 1)$, $j = 0, 1, \ldots, N$.

It is important to keep in mind that specification of the space $\mathscr{C}^k[a, b]$ and of a set of nodes does not guarantee the existence of a unique $f \in \mathscr{C}^k[a, b]$ having the prescribed nodal values. To illustrate the lack of uniqueness, we need only recall that the simple interpolation problem could be solved by polynomials, linear splines, or natural cubic splines, for example. Once we have chosen a family of functions in $\mathscr{C}^k[a, b]$, which, for a given set of nodes, yield a unique interpolant, the idea of nodal value is useful in obtaining representations of the interpolant in terms of cardinal functions. Keeping in mind that two ordered pairs $(u; i)$, $(v; j)$ are equal if and only if $u = v$ and $i = j$, the following assertion holds.

Theorem 8.4.1   *Let a set of $m + 1$ distinct nodes M_i, $i = 0, 1, \ldots, m$, be given (each being an ordered pair as in the definition), involving at most a kth derivative. Let $\mathscr{F}$ be a family of functions in $\mathscr{C}^k[a, b]$ with the property that there exists a unique f in $\mathscr{F}$ having the given nodal values. Then there exist $m + 1$ unique (cardinal) functions φ_i, $i = 0, 1, \ldots, m$, having the properties*:

$$\text{nodal value of } \varphi_i \text{ at } M_j = \begin{cases} 1, & i = j, \\ 0, & i \neq j. \end{cases}$$

Furthermore, if the given nodal values of f at M_i is denoted by v_i, then the interpolant has the representation

$$f(x) = \sum_{i=0}^{m} v_i \varphi_i(x).$$

We have seen explicit examples of such cardinal functions and associated representations of f in the previous chapters.

It is now a comparatively simple matter to extend the idea of "node" to two-dimensional problems on rectangular regions and to see how one-dimensional schemes can be combined to find interpolants on a rectangular lattice. For this purpose it is useful to have a formal definition of a "node" in the context of bivariate interpolation.

Definition   *If a function f defined on $[a, b] \times [c, d]$ is known to belong to the class $\mathscr{C}^{k+l}(S)$, where $S = [a, b] \times [c, d]$, and if the value of $\partial^{i+j}f(u, v)/\partial x^i \partial y^j$ is given for some i, j satisfying $0 \leq i \leq k$, $0 \leq j \leq l$, with $(u, v) \in [a, b] \times [c, d]$, then the ordered 4-tuple $(u, v; i, j)$ is said to be a node. We call the value of the partial derivative $\partial^{i+j}f(u, v)/\partial x^i \partial y^j$ the nodal value of f at this node.*

With reference to schemes of Section 8.3, we observe that if an interpolation scheme along the x-axis has nodes located at $x_0, x_1, \ldots, x_M$, and another scheme along the y-axis has nodes at $y_0, y_1, \ldots, y_N$, then the product scheme has nodes at the points (x_i, y_j). This tells us where the nodes are but not what they are. The following tells us how to construct nodes for more general product schemes and how we may get a representation for an interpolating function.

Theorem 8.4.2   *Let a set of distinct nodes M_i, $i = 0, 1, \ldots, m$, located in $[a, b]$ be given, involving at most a kth derivative. Let $\mathscr{F}$ be a family of functions in $\mathscr{C}^k[a, b]$ with the property that there exists a unique $f \in \mathscr{F}$ having the given nodal values. Let a second set of distinct nodes N_j, $j = 0, 1, \ldots, n$, located in $[c, d]$ be given, involving at most an lth derivative, and let $\mathscr{G}$ be a family of functions in $C^l[a, b]$ such that a unique $g \in \mathscr{G}$ exists having the given nodal values. Then there exists a unique interpolant P on $[a, b] \times [c, d]$ whose nodes are 4-tuples P_{ij}, $i = 0, 1, \ldots, m$, $j = 0, 1, \ldots, n$, obtained by combining all the nodes M_i with all the nodes N_j as follows: If $M_i = (u; r)$ and $N_j = (v; s)$, then*

$$P_{ij} = (u, v; r, s).$$

The nodal value of P at the nodes P_{ij} is then $\partial^{r+s}P(u, v)/\partial x^r \partial y^s$.

In addition, if the cardinal functions for the interpolation from $\mathscr{F}$ are $\varphi_0, \ldots, \varphi_m$ and those from $\mathscr{G}$ are $\psi_0, \ldots, \psi_n$ and if the nodal values of P at P_{ij} are denoted by v_{ij}, then a representation for P is

$$P(x, y) = \sum_{i=0}^{m} \sum_{j=0}^{n} v_{ij} \varphi_i(x) \psi_j(y).$$

The $(m + 1)(n + 1)$ products $\varphi_i(x)\psi_j(y)$, $i = 0, 1, \ldots, m$, $j = 0, 1 \ldots, n$ are the cardinal functions for the interpolation problem with nodes P_{ij}.

For the purpose of illustration, Theorem 8.4.2 will now be applied to construct Hermitian interpolation schemes on a rectangle. Let the rectangle be $[x_0, x_M] \times [y_0, y_N]$. In both the x- and y-directions, Hermitial interpolation is to be used, as described in Section 2.4. Then for each $i = 0, 1, \ldots, M$ and $j = 0, 1, \ldots, N$, there are two nodes for linear interpolation, namely, $(x_i; 0)$, $(x_i; 1)$, and $(y_j; 0)$, $(y_j; 1)$, respectively. These combine to give four nodes at each point (x_i, y_j) for the product scheme:

$$(x_i, y_j; 0, 0), \qquad (x_i, y_j; 1, 0), \qquad (x_i, y_j; 0, 1), \qquad (x_i, y_j; 1, 1),$$

and they represent the values of f, f_x, f_y, and f_{xy}, respectively, evaluated at (x_i, y_i). Thus, there are $4(M + 1)(N + 1)$ nodes in all.

It is important to note that, although the one-dimensional Hermitian interpolant requires only first derivatives (slope) information, the product scheme requires some *second* derivatives at the lattice points.

For this osculatory interpolation there are $2(M + 1)$ and $2(N + 1)$ cardinal basis functions $\varphi_i(x)$ and $\psi_j(y)$, respectively, associated with interpolation in the two coordinate directions. Hence, there are $4(M + 1)(N + 1)$ bivariate cardinal functions $\varphi_i(x)\psi_j(y)$, matching the total number of nodes. The simplest special case of this procedure will be, for us, the most important.

EXAMPLE 3   Consider the case $M = N = 1$ of the above discussion. Now the rectangle of interpolation is $[x_0, x_1] \times [y_0, y_1]$. In the x-direction four nodes are given: $(x_0; 0)$, $(x_0; 1)$, $(x_1; 0)$, and $(x_1; 1)$. Construction of corresponding cubic cardinal functions and interpolants has been considered in detail in Section 2.4 (although, for convenience, on the interval $[0, 1]$) and there are, in particular, just four basis functions in each direction which combine to give 16 basis functions for interpolation on the rectangle, matching the total of 16 nodes, four at each vertex. This is precisely the *bicubic rectangle* discussed in Section 7.6 (see especially Fig. 7.12). In contrast to the discussion of Section 7.6 we now have a constructive approach to the formulation of a cardinal basis for this interpolation scheme.

EXAMPLE 4   Consider once more a general rectangular lattice of points on the rectangle $[x_0, x_M] \times [y_0, y_N]$. Instead of the one-dimensional osculatory interpolation proposed above, we consider the *piecewise* cubic interpolation discussed in detail in Section 3.7 and to be implemented in both the x- and y-directions.

It is not difficult to see that the local property (on subintervals) of the one-dimensional scheme is inherited by the product scheme (on subrectangles $[x_{i-1}, x_i] \times [y_{j-1}, y_j]$). Thus, the implementation of this product scheme leads to a "patchwork" of $(M + 1)$ $(N + 1)$ problems posed on elementary rectangles. On each rectangle a problem like that of Example 3 is to be solved, and the 16 nodes on the perimeter of the rectangle determine completely the bicubic interpolant on the interior of the rectangle.

A point which will be particularly important subsequently is, that the *one-dimensional $\mathscr{C}^1$ property of the piecewise cubic Hermite interpolant is* *inherited by the patchwork of bicubic rectangles.* Thus, the surface $z = P(x, y)$ defined over $[x_0, x_M] \times [y_0, y_N]$ is a $\mathscr{C}^1$-surface, as defined in Section 1.6. Examples of this application can be found in Chapter 9.

8.5 Blending-function methods

These methods seem to have originated with manufacturers of car bodies and have since found many other industrial applications, ranging from the design of ship hulls to thermodynamics. As in the preceding sections, we desire a smooth surface to fit some data, but in this case the data is given in the form of functions describing the complete cross-sections of the surface on a number of intersecting curves. For the present purposes we look at the simpler situation where the curves are straight lines intersecting at right angles, and the region over which we are working is a rectangle with sides parallel to the lines (as described in Section 8.1).

Suppose then that the data occurs on the rectangle $[x_0, x_M] \times [y_0, y_N]$. On the lines $x = x_0$, $x = x_1$, ..., $x = x_M$ in the xy-plane, y can vary. We suppose that we are given $M + 1$ functions $g_0(y)$, ..., $g_M(y)$ representing cross sections of a surface on these lines. These functions may be produced from data available along the lines by means of interpolation methods for functions of a single variable, as discussed in earlier chapters.

Similarly, we are given $N + 1$ functions $h_0(x)$, ..., $h_N(x)$ describing the cross sections of the surface along the lines $y = y_0$, $y = y_1$, ..., $y = y_N$.

It turns out that, by using cardinal functions as in the product schemes already discussed, we are able to produce a surface $z = B(x, y)$ over the whole rectangle $[x_0, x_M] \times [y_0, y_N]$, which has the specified cross sections. As in Section 8.2, let $\varphi_i(x)$, $i = 0, 1, ..., M$, and $\psi_j(y)$, $j = 0, 1, ..., N$, be the cardinal functions for two selected interpolation methods, one to be used in each direction. Consistency demands that the cross-section curves agree in value where an x-section crosses a y-section. This means that

$$g_i(y_j) = h_j(x_i), \qquad i = 0, 1, ..., M, \quad j = 0, 1, ..., N.$$

These common values occur at the grid points and were previously denoted by f_{ij}.

We shall first develop the blending method in the simple case of data given along the four edges of the unit square $[0, 1] \times [0, 1]$. The interpolation method we choose is linear because there are only two grid points (or lines) in each direction. Referring to Figs. 8.8 and 8.9, we see that in our case $y_0 = 0$, $y_1 = 1$, $x_0 = 0$, $x_1 = 1$, $N = M = 1$. The given profiles are $h_0(x)$, $h_1(x)$, $g_0(y)$, $g_1(y)$, and the consistency conditions are

$$g_i(y_j) = h_j(x_i) = f_{ij}, \qquad i, j = 0, 1. \tag{8.7}$$

The cardinal functions for linear interpolation in the two directions are [cf. Eqs. (8.5) and (8.6)]:

$$\varphi_0(x) = 1 - x, \qquad \varphi_1(x) = x, \quad \psi_0(y) = 1 - y, \qquad \psi_1(y) = y.$$

It is quite easy to find two surfaces which effect the transition from $h_0(x)$ to $h_1(x)$ and from $g_0(y)$ to $g_1(y)$. These are

$$H(x, y) = h_0(x)\psi_0(y) + h_1(x)\psi_1(y),$$
$$G(x, y) = g_0(y)\varphi_0(x) + g_1(y)\varphi_1(x),$$

respectively.

It is the cardinality properties of $\varphi_0(x)$, $\varphi_1(x)$, $\psi_0(y)$, $\psi_1(y)$ that ensure that

$$H(x, 0) = h_0(x), \qquad H(x, 1) = h_1(x),$$
$$G(0, y) = g_0(y), \qquad G(1, y) = g_1(y).$$

It turns out that the functions $H(x, y)$ and $G(x, y)$ can be combined with the product interpolant defined in Section 8.2, $P(x, y)$, to produce the desired function $B(x, y)$ that coincides with the four given profiles at the edges (Fig. 8.10). We have [see Eqs. (8.2) and (8.3)],

$$P(x, y) = f_{00}\varphi_0(x)\psi_0(y) + f_{10}\varphi_1(x)\psi_0(y) + f_{11}\varphi_1(x)\psi_1(y)$$
$$+ f_{01}\varphi_0(x)\psi_1(y),$$

and claim that

$$B(x, y) = H(x, y) + G(x, y) - P(x, y)$$

will do the trick. Certainly, this makes sense at the corners, where the consistency of the given profiles results in $H(x, y) + G(x, y)$ having double the required value, this being corrected by subtracting $P(x, y)$. Let us check along the line $y = 0$:

$$
\begin{aligned}
B(x, 0) &= H(x, 0) + G(x, 0) - P(x, 0) \\
&= h_0(x) + g_0(0)\varphi_0(x) + g_1(0)\varphi_1(x) - f_{00}\varphi_0(x)\psi_0(0) \\
&\quad - f_{10}\varphi_1(x)\psi_0(0) - f_{11}\varphi_1(x)\psi_1(0) - f_{01}\varphi_0(x)\psi_1(0) \\
&= h_0(x) + f_{00}\varphi_0(x) + f_{10}\varphi_1(x) - f_{00}\varphi_0(x) \cdot 1 \\
&\quad - f_{10}\varphi_1(x) \, 1 - f_{11}\varphi_1(x) \cdot 0 - f_{01}\varphi_0(x) \cdot 0 \\
&= h_0(x).
\end{aligned}
$$

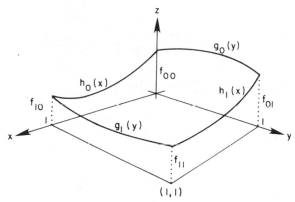

FIG. 8.8   Blending data on the unit square.

Here we have exploited the consistency conditions of Eq. (8.7) and the cardinality conditions satisfied by $\varphi_0(x)$, $\varphi_1(x)$, $\psi_0(y)$, $\psi_1(y)$. In a similar way we can verify that $B(x, 1) = h_1(x)$, $B(0, y) = g_0(y)$, and $B(1, y) = g_1(y)$, so that $B(x, y)$ performs as required. The function $B(x, y)$ is called a *linearly blended* interpolant because it blends together the functions given on the edges using linear interpolation in the two directions.

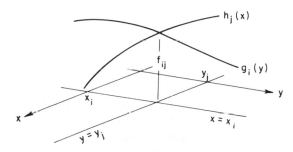

FIG. 8.9   Consistency condition.

The same construction can be employed in the general case with data and consistency conditions as in figs. 8.9 and 8.10. The general result is presented in a formal statement.

Theorem 8.5.1.   *Let a rectangular lattice be defined by the points $x_0 < x_1 < \ldots < x_M$ and $y_0 < y_1 < \ldots < y_N$. Let $M + 1$ functions g_i, $i = 0, 1, \ldots, M$, defined for $y_0 \leq y \leq y_N$, and $N + 1$ functions h_j, $j = 0, 1, \ldots, N$, defined for $x_0 \leq x \leq x_M$, be given, satisfying*

$$g_i(y_j) = h_j(x_i) = f_{ij}, \qquad i = 0, 1, \ldots, M, \quad j = 0, 1, \ldots, N.$$

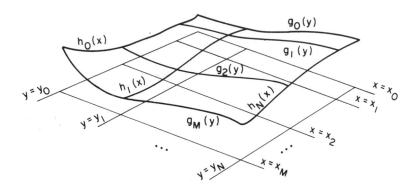

FIG. 8.10   Blending grid.

If $\varphi_0, \ldots, \varphi_M$ and $\psi_0, \ldots, \psi_N$ are cardinal functions for interpolation at the nodes $(x_0; 0), \ldots, (x_M; 0)$ and $(y_0; 0), \ldots, (y_N; 0)$ (in the notation of Section 8.4), respectively, then the blended interpolant $B(x, y)$ defined by

$$B(x, y) = \sum_{i=0}^{M} \varphi_i(x)g_i(y) + \sum_{j=0}^{N} h_j(x)\psi_j(y) - \sum_{i=0}^{M}\sum_{j=0}^{N} f_{ij}\varphi_i(x)\psi_j(y), \quad (8.8)$$

has the properties

$$B(x_i, y) = g_i(y), \quad i = 0, 1, \ldots, M, \qquad B(x, y_j) = h_j(x), \quad j = 0, 1, \ldots, N.$$

In applications, the functions g_i and h_j will usually not be available, and will have to be constructed by using interpolation or smoothing methods. These methods need not be the same as the methods implied by the cardinal functions φ_i and ψ_j being used for blending.

An important distinction between product interpolants and blended interpolants is that whereas the former uses only a finite number of nodes, the latter utilizes information along entire lines which are formed from infinitely many points.

Theorem 8.5.1 concerns itself with blending information about function values along lines and does not utilize derivative information. However, a more general statement can be made that provides for blending with, for example, the cardinal functions for piecewise cubic Hermitian interpolation.

Theorem 8.5.2   *Let all the conditions and notation be as in Theorem* 8.4.2. *In addition, assume that the nodes M_i, $i = 0, 1, \ldots, m$, $m \geq M$ are located at the points $x_0 < x_1 < \ldots < x_M$ (several nodes can have the same location),* and that the nodes N_j, $j = 0, 1, \ldots, n$, $n \geqslant N$ are located at $y_0 < y_1 < \ldots < y_N$. Let a given function f defined on $[x_0, x_M] \times [y_0, y_N]$ be continuously differentiable k times with respect to x and l times with respect to y.

If a node M_i has the form $(u; r)$, define the nodal value of f at this node by

$$g_i(y) = \partial^r f(u, y)/\partial x^r, \qquad i = 0, 1, \ldots, m.$$

Similarly, if a node N_j has the form $(v; s)$, define

$$h_j(x) = \partial^r f(x, v)/\partial y^s, \qquad j = 0, 1, \ldots, n.$$

The blended interpolant of f is then given by

$$B(x, y) = \sum_{i=0}^{m} \varphi_i(x)g_i(y) + \sum_{j=0}^{n} h_j(x)\psi_j(y) - \sum_{i=0}^{m}\sum_{j=0}^{n} v_{ij}\varphi_i(x)\psi_j(y),$$

and has the properties

$$\partial^r B(u, y)/\partial x^r = g_i(y), \quad i = 0, 1, \ldots, m;$$

$$\partial^s B(x, v)/\partial y^s = h_j(x), \quad j = 0, 1, \ldots, n.$$

Note that in this statement, by expressing the blended interpolant in terms of nodal values of an underlying function f, we have automatically taken care of the kind of consistency conditions that appear in Theorem 8.5.1 in the form $g_i(y_j) = h_j(x_i)$. Also, recall that v_{ij} is the nodal value of f at node P_{ij} (see Theorem 8.4.2). If, as will often be the case in practice, such a function is not available, consistency conditions will have to be imposed. These are too complicated to express here in full generality, and we will content ourselves with an example.

EXAMPLE *Hermitian Blending.* We select the region $[0, 1] \times [0, 1]$ with the lattice defined by $x_0 = 0$, $x_1 = 1$, $y_0 = 0$, $y_1 = 1$. The nodes are the same as in the example of Section 8.4, so $m = n = 3$, whereas $M = N = 1$. Now,

$$M_0 = (0; 0) \Rightarrow g_0(y) = f(0, y), \qquad N_0 = (0; 0) \Rightarrow h_0(x) = f(x, 0),$$

$$M_1 = (0; 1) \Rightarrow g_1(y) = \frac{\partial f(0, y)}{\partial x}, \qquad N_1 = (0; 1) \Rightarrow h_1(x) = \frac{\partial f(x, 0)}{\partial y},$$

$$M_2 = (1; 0) \Rightarrow g_2(y) = f(1, y), \qquad N_2 = (1; 0) \Rightarrow h_2(x) = f(x, 1),$$

$$M_3 = (1; 1) \Rightarrow g_3(y) = \frac{\partial f(1, y)}{\partial x}, \qquad N_3 = (1; 1) \Rightarrow h_3(x) = \frac{\partial f(x, 1)}{\partial y}.$$

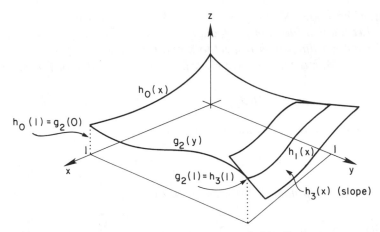

FIG. 8.11   Consistency conditions in blending.

The presence of the functions g_1 and g_3 expresses the fact that the slope of the surface in the x-direction is to be given when progressing along the lines $x = 0$ and $x = 1$ in the y-direction. A similar statement applies with respect to the nodes N_1, N_3 and corresponding functions h_1, h_3. If the function f is not given, then g_0, g_2, h_0, h_2 must be obtained by interpolating or smoothing whatever data are available along the lines $x = 0$, $x = 1$, $y = 0$, $y = 1$. The transverse slopes in both the x- and y-directions will have

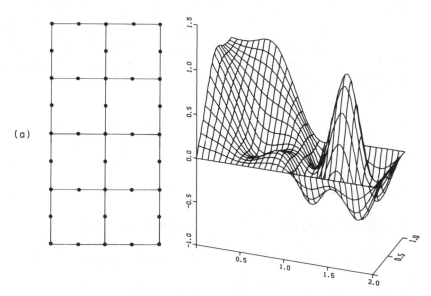

FIG. 8.12   Three spline blended surface interpolants.

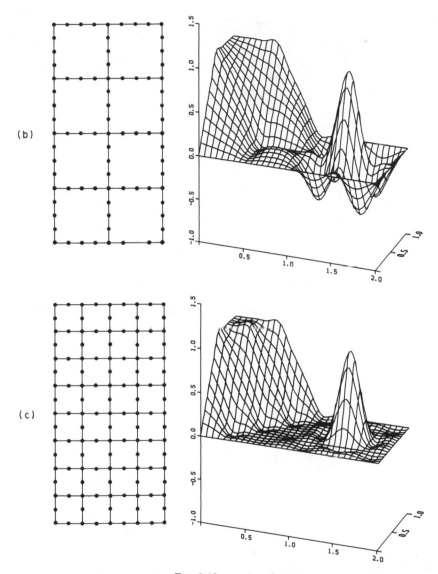

FIG. 8.12 *continued.*

to be estimated and smoothed or interpolated to yield g_1, g_3, h_1, h_3. The functions g_i and h_j are not independent of each other, for they must conform to an underlying continuously differentiable function. Thus, we impose the conditions

$$g_0(0) = h_0(0), \qquad g_0(1) = h_2(0), \qquad g_2(0) = h_0(1), \qquad g_2(1) = h_2(1),$$

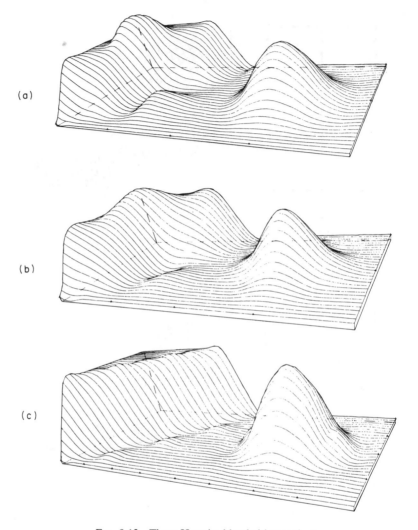

(a)

(b)

(c)

FIG. 8.13   Three Hermite blended interpolants.

in order that the cross-sectional profiles agree at the corners. Further, the transverse slope along an edge must agree with the slope of the tangent to the profile encountered at the corners:

$$g_1(0) = h'_0(0), \qquad h_1(0) = g'_0(1), \qquad g_1(1) = h'_2(0), \qquad h_1(1) = g'_2(1),$$
$$g_3(0) = h'_0(1), \qquad h_3(0) = g'_0(1), \qquad g_3(1) = h'_2(1), \qquad h_3(1) = g'_2(1).$$

Some of these are illustrated in Fig. 8.11.

In Fig. 8.12 we illustrate the performance of the cubic spline blending technique on the model problem described in Section 7.7. The improved resolution with the increasing numbers of data points is clear. Irregularities, which appear with a wavelength determined by the underlying grid, are also apparent. The points on the grid lines alongside the figures of 8.12 show where data were supplied. The cubic spline required only elevations at the points to produce the interpolants $g_i(y)$ and $h_j(x)$; These were then blended using cardinal spline functions based on the coarser grid line spacing. The surface of Fig. 8.12c has larger oscillations than the bicubic spline interpolant of Fig. 8.5c; however, it, unlike the bicubic spline, does fit the data along entire lines. Another example of spline blending occurs in Section 9.11.

The surfaces of Fig. 8.13 were generated by a Hermitian blending technique. In this case, transverse slopes were supplied as well as elevations. For interpolation along the lines, slopes were approximated as at the end of Section 3.2. The transverse slope was interpolated in the same way, and thus the blended surface has close to correct elevations and transverse slopes on all grid lines. A contour map of Fig. 8.13c is presented in Fig.

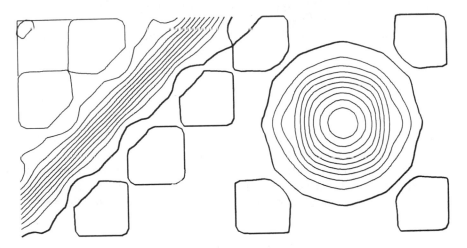

FIG. 8.14   Contour map of the surface in Fig. 8.13c.

8.14. Comparison of Figs. 8.12 and 8.14 suggests that the Hermitian technique has not produced a noticeably better resolution of the model surface, even though it requires considerably more programming and computational effort.

All of the interpolation and blending methods introduce anomalies that were certainly not in the original surface. Part of the problem is that the original surface is only of class $\mathscr{C}$ because of the sharp corners of the ramp, whereas all of the approximating surfaces are smooth, at least to the extent of having continuous slopes; they are of class $\mathscr{C}^1$ and cannot fit sharp corners. More details and references on blending function methods can be found in papers by W. J. Gordon (1968; 1969a, b), for example.

8.6 Product and blending schemes as projection methods

We now indicate how the language of projectors, as introduced in Chapter 5, can be used in the discussion and analysis of product and blending methods. This formalism becomes particularly useful and elegant in this context. For simplicity we confine discussion to the interpolation problems in which only function value data is given at the vertices of our standard rectangular lattice in the case of product schemes and at all points of the edges of the lattice in the case of blending.

Recall first the projector associated with interpolation to function values $f(x_0), \ldots, f(x_M)$ at points $x_0, x_1, \ldots, x_M$ on a line and, in particular, the meaning of Eq. (5.7). We suppose that we have a well-defined interpolation scheme with corresponding cardinal functions $\varphi_0(x), \ldots, \varphi_M(x)$. Thus, for a given function $f(x) \in \mathscr{C}[a, b]$, say, there is a unique interpolant in the set $\mathscr{T}_1$ of linear combinations of $\varphi_0(x), \ldots, \varphi_N(x)$ given by

$$(P_0 f)(x) = \sum_{i=0}^{M} \varphi_i(x) f(x_i). \tag{8.9}$$

Equation (8.9) defines the map P_0 from functions f in $\mathscr{C}[a, b]$ to functions in $\mathscr{T}_1$, and P_0 is a projector in the sense that $P_0^2 = P_0$.

We now wish to extend the definition of P_0 to apply to functions $f(x, y)$ that are continuous on $[x_0, x_M] \times [y_0, y_N]$. This can be done formally by simply introducing y as a "sleeping partner" in Eq. (8.9). Thus, Eq. (8.10),

$$(Pf)(x, y) = \sum_{i=0}^{M} \varphi_i(x) f(x_i, y), \tag{8.10}$$

determines the action of P on horizontal lines in one coordinate plane determined by fixed values of y. This defines the image of P as being functions that for each fixed y are in $\mathscr{T}_1$ but for each fixed x are merely continuous functions on $[y_0, y_N]$.

Now introduce an interpolation scheme with respect to the y-variable and determined by cardinal functions $\psi_0(y), \psi_1(y), \ldots, \psi_N(y)$. Then, in just the same manner, we construct a map Q by means of the relation

$$(Qf)(x, y) = \sum_{j=0}^{N} \psi_j(y)f(x, y_j), \qquad (8.11)$$

and functions in the image of Q will be continuous with respect to x for each fixed y, and for each fixed x they are functions in the class $\mathcal{T}_2$ of linear combinations of $\psi_0(y), \ldots, \psi_N(y)$. It is easily verified that the maps P and Q constructed in this way are indeed projectors. That is, they satisfy $P^2 = P$ and $Q^2 = Q$.

Now consider the function of x and y given by Eq. (8.11) and apply the projector P to this function. It is found that

$$(PQf)(x, y) = \sum_{i=0}^{N} \varphi_i(x) \sum_{j=0}^{M} \psi_j(y)f(x_i, y_j) = \sum_{j=0}^{N} \sum_{i=0}^{M} f_{ij} \varphi_i(x)\psi_j(y). \qquad (8.12)$$

However, comparing with Eqs. (8.3) and (8.1), we see that this is just the interpolant of the product scheme discussed earlier.

On applying the projector Q to both sides of Eq. (8.10), it is found that

$$(QPf)(x, y) = \sum_{j=0}^{N} \sum_{i=0}^{M} f_{ij} \varphi_i(x)\psi_j(y),$$

as well. Thus, the projectors P and Q *commute*; we have $PQ = QP$. Let us summarize our findings: *The maps P and Q defined by Eqs. (8.10) and (8.11) are commuting projectors and, for any function f defined at the vertices of the lattice, the function $QPf = PQf$ is the interpolant to f at the vertices determined by the product scheme with cardinal functions $\varphi_i(x)\psi_j(y)$ for i = 0, 1, \ldots, M and j = 0, 1, \ldots, N. Furthermore, the map PQ is itself a projector.*

The last statement follows from the commutativity of P and Q. For $(PQ)^2 = P(QP)Q = P^2Q^2 = PQ$. This formulation simplifies the analysis of product interpolation schemes enormously and admits a geometric interpretation of the associated projector PQ.

Now consider Eq. (8.8), giving an explicit formula for a blended interpolant to the data $f(x_i, y) = g_i(y)$ on the lines $x = x_i$ $(i = 0, 1, \ldots, M)$ and $f(x, y_j) = h_j(x)$ on the lines $y = y_j$ $(j = 0, 1, \ldots, N)$. Comparing this with Eqs. (8.10)–(8.12), we see at once that the blended interpolant can be written in terms of the projectors P and Q:

$$B(x, y) = (Pf)(x, y) + (Qf)(x, y) - (PQf)(x, y),$$

or, more concisely, the blended interpolant to $f(x, y)$ is obtained by applying the map $P + Q - PQ$ to f. This composite map is the *Boolean sum* of P and Q and is written $P \oplus Q$. (Compare with Section 5.4. There, the situation is significantly different because the projectors P and Q do not commute.)

It is easily verified that, once more, $P \oplus Q$ is a projector and useful properties of $P \oplus Q$, and hence blending interpolation, can be derived directly from properties of the separate projectors P and Q.

For the benefit of readers familiar with the ideas of the intersection and sum of subspaces, we record the following useful facts. If P and Q are any commuting projectors, then

$$\text{Im } PQ = (\text{Im } P) \cap (\text{Im } Q), \qquad \text{Ker } PQ = (\text{Ker } P) + (\text{Ker } Q),$$

$$\text{Im}(P \oplus Q) = (\text{Im } P) + (\text{Im } Q), \qquad \text{Ker}(P \oplus Q) = (\text{Ker } P) \cap (\text{Ker } Q).$$

9

Finite Element Methods

9.1 Introduction

Our purpose in this chapter is to describe the use of finite element techniques for surface construction, and, in the main, we shall develop the exposition through successively more difficult problems. Indeed, for the purpose of this introduction, we can use techniques of interpolation *on a line* to develop the underlying ideas in a simple way.

First, recall the technique of interpolation on a line with the use of piecewise linear functions, as described in Sections 3.2 and 3.3. In the notation of Theorem 3.3.1, distinct abscissas $x_0 < x_1 < \ldots < x_N$ are given and function values of $f_0, f_1, \ldots, f_N$ are assigned. In contrast to the "global" view of Theorem 3.3.1, the interpolation can obviously proceed "locally", sub-interval by sub-interval. On each sub-interval $[x_j, x_{j+1}]$, $j = 0, 1, \ldots, N - 1$, we can construct a unique linear interpolant,

$$l_j(x) = [(f_{j+1} - f_j)/(x_{j+1} - x_j)](x - x_j) + f_j, \tag{9.1}$$

which we apply only to those values of x in $[x_j, x_{j+1}]$. By "abutting" these interpolants and sub-intervals to one another, the unique interpolant $l(x)$ on $[x_0, x_N]$ (whose existence is asserted in Theorem 3.3.1) is generated.

In this construction the components of each interpolation problem on a sub-interval determine a *finite element*. Thus, a typical finite element consists of the following three parts:

(1) an interval $[x_j, x_{j+1}]$;
(2) the nodes $(x_j; 0)$ and $(x_{j+1}; 0)$ (in the notation of Section 8.4); and
(3) the class of functions $\mathscr{P}_1$.

Now it seems trivial in this context, but the function of Eq. (9.1) for the jth sub-interval can be generated by first interpolating on the "standard" interval $[0, 1]$ and then transforming to $[x_j, x_{j+1}]$. Note first that if we

179

introduce a new variable ξ by writing

$$\xi = (x - x_j)/(x_{j+1} - x_j), \tag{9.2}$$

then the interval $[x_j, x_{j+1}]$ on the x-axis maps onto the interval $[0, 1]$ of the ξ-axis.

For linear interpolation on $[0, 1]$ we have the simple cardinal functions

$$\varphi_0(\xi) = 1 - \xi, \qquad \varphi_1(\xi) = \xi,$$

which immediately tell us that the linear interpolant to the values f_j, f_{j+1} at 0 and 1, respectively, is

$$\hat{l}_j(\xi) = f_j(1 - \xi) + f_{j+1}\xi = (f_{j+1} - f_j)\xi + f_j.$$

To recover the interpolant [Eq. (9.1)] on $[x_j, x_{j+1}]$, we simply substitute for ξ from Eq. (9.2).

The purpose of this manoeuvre is to replace interpolation on N different finite elements by N interpolation problems on the same standard element, using the same cardinal functions in each case, and followed by the coordinate transformation of Eq. (9.2). The "standard" finite element for this procedure is, of course, determined by

(1)   the interval $[0, 1]$;
(2)   the nodes $(0; 0)$ and $(1; 0)$; and
(3)   the class of functions $\mathcal{P}_1$.

Let us briefly indicate how the same ideas apply to the piecewise cubic interpolation (on the line) of Sections 3.6 and 3.7. It is supposed that abscissas and function values are assigned as above and, in addition, the slopes of the interpolating function $m_0, m_1, \ldots, m_N$ (at $x_0, x_1, \ldots, x_N$, respectively) are assigned. A cubic function $S_j(x)$ is to be determined which interpolates the data f_j, m_j at x_j and f_{j+1}, m_{j+1} at x_{j+1}, this for $j = 0, 1, \ldots, N - 1$. Again, the problem can be handled sub-interval by sub-interval. However, as with the piecewise linear functions we choose to perform all the interpolation on a standard sub-interval. In this case the standard finite element consists of

(1)   the interval $[0, 1]$;
(2)   the nodes $(0; 0)$, $(0; 1)$, $(1; 0)$, and $(1; 1)$; and
(3)   the class of functions $\mathcal{P}_3$.

The cardinal functions are, respectively [see Eqs. (2.9), (3.14), and (3.15)],

$$H_0(\xi) = (2\xi + 1)(\xi - 1)^2, \qquad K_0(\xi) = \xi(\xi - 1)^2,$$
$$H_1(\xi) = \xi^2(3 - 2\xi), \qquad K_1(\xi) = \xi^2(\xi - 1).$$

To construct the interpolant on $[x_j, x_{j+1}]$, we first construct a corresponding interpolant on the standard element $[0, 1]$. Compared to the previous example involving a piecewise linear interpolant, a new feature enters at this point. The function value data f_j and f_{j+1} can simply be carried over to the standard element, but the slope data m_j, m_{j+1} must be adjusted to account for the change in scale in going from one element to the other. We define the transformed data values

$$\hat{m}_j = (x_{j+1} - x_j)m_j, \qquad \hat{m}_{j+1} = (x_{j+1} - x_j)m_{j+1},$$

and apply the data $f_j, \hat{m}_j, f_{j+1}, \hat{m}_{j+1}$ on the standard element to construct the interpolant

$$\hat{S}_j(\xi) = f_j H_0(\xi) + \hat{m}_j K_0(\xi) + f_{j+1} H_1(\xi) + \hat{m}_{j+1} K_1(\xi),$$

where $\xi \in [0, 1]$. Then the interpolant on $[x_j, x_{j+1}]$ is determined by means of Eq. (8.2):

$$S_j(x) = \hat{S}_j[(x - x_j)/(x_{j+1} - x_j)].$$

On repeating this process for $j = 0, 1, \ldots, N - 1$, the complete interpolant on $[x_0, x_N]$ is obtained.

The point to be emphasized here and carried to interpolation in the plane is that a finite element consists of a geometrical figure together with an interpolation scheme (consisting of a set of functions and a set of nodes) defined on that figure. Furthermore, if possible, interpolation is to be performed on a "standard" geometrical figure of simple form, and interpolants on more general figures are to be determined by (affine) transformations.

In these two examples the standard "geometrical figure" is the same, but the prescribed interpolation schemes are different. In the first case the result is a $\mathscr{C}^0$ interpolant on $[x_0, x_N]$, and there is continuity of the function values at $x_1, x_2, \ldots, x_{N-1}$ but not (in general) of the derivatives. In the second case, by including derivative values as nodes at $x_1, x_2, \ldots, x_{N-1}$, an interpolating curve of class $\mathscr{C}^1$ is guaranteed.

For surface construction our presentation will be confined to rectangular and triangular "geometrical figures". The region of the plane on which the surface is to be constructed will be filled with rectangles or triangles and an interpolation scheme associated with each one. Recall that several ideas relevant to this program have already been developed in Chapter 7; the reader should be sure to have the material of that chapter firmly in mind before proceeding. We draw attention to the interpolation schemes on rectangles of Section 7.6 and the associated transformations of Eq. (7.4). The reader should recall as well the construction of interpolation schemes on triangles of Section 7.5 and the associated affine transformations of Eqs. (7.9) and (7.10).

9.2 The bilinear rectangle

We begin with a discussion of a finite element scheme in the plane based on Theorem 7.4.1. Suppose that (as in Section 8.1) a surface is to be constructed on a rectangular domain $[x_0, x_M] \times [y_0, y_N]$ and that this is broken down into sub-rectangles $[x_j, x_{j+1}] \times [y_k, y_{k+1}]$, where $j = 0, 1, \ldots, M - 1$ and $k = 0, 1, \ldots, N - 1$, $x_0 < x_1 < x_2 < \ldots < x_M$ and $y_0 < y_1 < \ldots < y_N$. It is assumed that function values f_{jk} are given at each of the $(M + 1)(N + 1)$ vertices (x_j, y_k). Thus, Theorem 7.4.1 may be applied to each sub-rectangle and the data used to construct a bilinear interpolant defined on the *interior* of each sub-rectangle by the data at the four vertices of the same sub-rectangle. In this way we can construct an interpolant over the whole rectangle $[x_0, x_M] \times [y_0, y_N]$ as a patchwork of the separate bilinear interpolants defined one on each sub-rectangle.

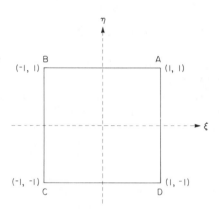

FIG. 9.1   Standard "rectangle".

Thus, in this technique a finite element consists of

(1)   a sub-rectangle $[x_j, x_{j+1}] \times [y_k, y_{k+1}]$,
(2)   the nodes (see the definition of Section 8.4)

$$(x_j, y_k; 0, 0), \qquad (x_{j+1}, y_k; 0, 0),$$

$$(x_{j+1}, y_{k+1}; 0, 0), \qquad (x_j, y_{k+1}; 0, 0),$$

(3)   The class of bilinear functions.

(Recall that the class of bilinear functions is spanned by the monomials 1, x, y, and xy.) For the implementation of the technique we reintroduce (Fig. 9.1) the standard "rectangle" $[-1, 1] \times [-1, 1]$ (see also Fig. 7.5). There is a corresponding standard finite element in which $x_j, x_{j+1}, y_k, y_{k+1}$ are replaced in the above definition by $-1, 1, -1$, and 1, respectively.

Then, in numerical practice all the interpolation is completed on this standard rectangle. The corresponding functions are

$$\varphi_A(\xi, \eta) = \tfrac{1}{4}(1 + \xi + \eta + \xi\eta), \qquad \varphi_B(\xi, \eta) = \tfrac{1}{4}(1 - \xi + \eta - \xi\eta),$$
$$\varphi_C(\xi, \eta) = \tfrac{1}{4}(1 - \xi - \eta + \xi\eta), \qquad \varphi_D(\xi, \eta) = \tfrac{1}{4}(1 + \xi - \eta - \xi\eta). \tag{9.3}$$

Here, $\varphi_A(\xi, \eta)$ has function value 1 at vertex A and 0 at the other vertices, and so on.

When working on the sub-rectangle $[x_j, x_{j+1}] \times [y_k, y_{k+1}]$, the interpolant to data $f_{j+1,k+1}, f_{j,k+1}, f_{j,k}, f_{j+1,k}$ at vertices A, B, C, D, respectively, is first constructed in the form

$$\hat{p}_{jk}(\xi, \eta) = f_{j+1,k+1}\varphi_A(\xi, \eta) + f_{j,k+1}\varphi_B(\xi, \eta) + f_{j,k}\varphi_C(\xi, \eta)$$
$$+ f_{j+1,k}\varphi_D(\xi, \eta). \tag{9.4}$$

The bilinear interpolant on $[x_j, x_{j+1}] \times [y_k, y_{k+1}]$ itself is then formed using the transformation [see Eq. (7.4)],

$$\xi = 1 - \frac{2(x_{j+1} - x)}{x_{j+1} - x_j}, \qquad \eta = 1 - \frac{2(y_{j+1} - y)}{y_{j+1} - y_j}. \tag{9.5}$$

That is, these expressions are simply substituted in Eq. (9.4) to obtain the desired interpolant

$$p_{jk}(x, y) = \hat{p}_{jk}(\xi, \eta).$$

This completes our description of the method for generating an interpolating surface on $[x_0, x_M] \times [y_0, y_N]$ using the bilinear finite element (see also Examples 1 and 3 of Section 8.3); however, what can we say about the properties of the surface generated in this way? First, it represents a function of class $\mathscr{C}^0$ — a function which is continuous on the whole rectangle. It is obviously continuous on the interior of any sub-rectangle, so the only question arises at an edge common to two sub-rectangles. For example, consider the edge joining points (x_j, y_k) and (x_j, y_{k+1}), as in Fig. 9.2, in

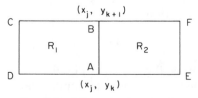

FIG. 9.2 Adjacent sub-rectangles.

which the adjacent rectangles are denoted by R_1 and R_2. Let the interpolating functions in these two rectangles be $p_1(x, y)$ and $p_2(x, y)$, respectively. Then, they have the forms

$$p_1(x, y) = a_1 + b_1 x + c_1 y + d_1 xy, \qquad p_2(x, y) = a_2 + b_2 x + c_2 y + d_2 xy,$$

for some constants $a_1, b_1, c_1, d_1, a_2, b_2, c_2, d_2$.

Why should these two functions take the same value on the edge common to R_1 and R_2? For this is what $\mathscr{C}^0$ continuity of the whole patchwork surface implies. The reason is that when *restricted* to this common edge, both $p_1(x_j, y)$ and $p_2(x_j, y)$ are simply linear functions of y. Since they take common values f_{jk} and $f_{j,k+1}$ at (x_j, y_k) and (x_j, y_{k+1}), respectively, they must be the *same* linear function, i.e. $p_1(x_j, y) = p_2(x_j, y)$ for all y. This argument applies to every edge between two sub-rectangles and shows that, indeed, a $\mathscr{C}^0$ surface is obtained. However, it is easily seen that, in general, the whole surface is *not* of class $\mathscr{C}^1$. There will generally be "creases" along the edges common to two subrectangles. For example, suppose that function values 1 are assigned to vertices A and B in Fig. 9.2 and the value 0 is assigned to the other four. Then the interpolant on each rectangle is simply a plane, and the two planes join (in a continuous way) along AB. The interpolating surface has the form of a "ridge-tent" with a crease along AB.

For many practical purposes, the fact that the surfaces generated by the bilinear rectangle are not generally $\mathscr{C}^1$ will disqualify this procedure. If we insist on using rectangles *and* generating a $\mathscr{C}^1$ surface, it is necessary to consider more sophisticated interpolation schemes on rectangles, with a view to increasing the interelement continuity. Thus, although the bilinear rectangle may be useful in some cases, its inclusion here is mainly for expository purposes.

In principle, the higher order interpolation schemes summarized in Fig. 7.10 can also be used as bases for finite element schemes, but like the bilinear rectangle, they will not generally determine a surface with continuous transverse derivatives across element boundaries. (The scheme associated with the nodes of Fig. 7.10b is widely used in the solution of partial differential equations and is known as the Adini rectangle.)

Let us introduce an illustrative example that will be used repeatedly in the subsequent sections. Let a function $f(x, y)$ be defined on the square $[1, 2] \times [1, 2]$ by the Eq. (9.6):

$$z = 10(e^{2x} - 1) \sin \pi(y - 1) - 15(y - 3/2)(y - 5/2)(x - 1), \quad (9.6)$$

The surface $z = f(x, y)$ is illustrated in Fig. 9.3.

To illustrate the use of rectangular finite elements, we suppose the domain to be split into two parts, as indicated in Fig. 9.4. For rectangle 1

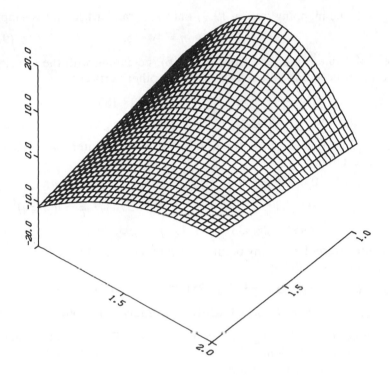

FIG. 9.3 An illustrative example.

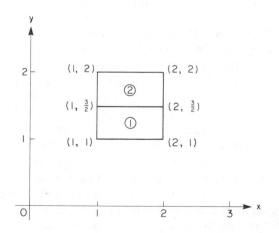

FIG. 9.4 Two rectangular elements.

in Fig. 9.4, the mapping by Eq. (9.9) onto the standard square is simply,

$$\xi = 2x - 3, \qquad \eta = 4y - 5. \tag{9.7}$$

The nodal values are read from Eq. (9.6), beginning with the top right vertex and proceeding anti-clockwise to the other vertices:

$$f(2, 3/2) = 0.0, \qquad f(1, 3/2) = 17.183,$$

$$f(1, 1) = 0.0, \qquad f(2, 1) = -11.25.$$

Using Eqs. (9.3) and (9.4), we obtain the bilinear interpolant on the standard rectangle:

$$4\hat{p}(\xi, \eta) = 0.0(1 + \xi + \eta + \xi\eta) + 17.183(1 - \xi + \eta - \xi\eta)$$

$$+ 0.0(1 - \xi - \eta + \xi\eta) + (-11.25)(1 + \xi - \eta - \xi\eta)$$

$$= 5.933 - 28.433\xi + 28.433\eta - 5.933\xi\eta.$$

Finally, the interpolant at any point (x_0, y_0) in rectangle 1 (Fig. 9.4) is now found by using Eq. (9.7):

$$p(x_0, y_0) = \hat{p}(2x_0 - 3, 4y_0 - 5)$$

$$= -34.982 + 0.616x_0 + 46.232y_0 - 11.866x_0y_0.$$

The computation on rectangle 1 is completed by taking as many values of (x_0, y_0) as required to define the surface adequately and by evaluating the last expression on each occasion.

Starting with the coordinates of the vertices of rectangle 2, the procedure can be repeated, beginning with the reformulation of Eq. (9.7). The result of these calculations on rectangles 1 and 2, is illustrated in Fig. 9.5. The discontinuity in the first derivative along the common edge of rectangles 1 and 2 is clear and, on comparison with Fig. 9.3, it is clear that for most purposes we do not have a useful approximation to the original surface.

If we are to keep the ridge from appearing in the approximation of Fig. 9.5 to the smooth surface of Fig. 9.3 (and retain polynomial interpolation on rectangles 1 and 2 of Fig. 9.4), it is necessary to examine interpolants of higher degree—those having the ability to ensure continuity of the first derivatives across the common edge of rectangles 1 and 2. On examining Section 7.6, the quadratic and cubic schemes of Figs. 7.10 and 7.11 suggest themselves. Unfortunately, neither of these have the property that this degree of continuity on the common edge can be guaranteed. The scheme of Fig. 7.11b gives a unique first derivative at each of the common vertices but not at points in the interior of the common edge. This Adini rectangle is not adequate for our purposes. It is necessary to tackle the complications of the "bicubic rectangle" of Fig. 7.12 before attaining the necessary interelement continuity.

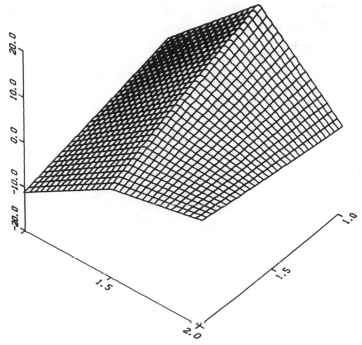

FIG. 9.5 Two bilinear rectangles and the model problem.

9.3 The bicubic rectangle

We have discussed a particular interpolation scheme in Chapters 7 and 8 known as the bicubic rectangle. The point of view of Chapter 7 (Fig. 7.12) was simply that a certain specification of 16 nodes and 16 monomial functions generate a well-defined interpolation scheme. In Chapter 8 (Examples 3 and 4 of Section 8.4), it was shown that this interpolation can be generated as a tensor product scheme based on hermite interpolation on the line, as developed in detail in Section 2.4.

Now we observe that in the context of Section 9.2, the bicubic rectangle can be used as a finite element. Furthermore, as Example 4 of Section 8.4 shows, *the surface generated by the use of this finite element will be of class* $\mathscr{C}^1$. This gives the bicubic rectangle an important place in an armory of finite element techniques. However, the $\mathscr{C}^1$ continuity is bought at the expense of introducing the *second*-order derivatives $\partial^2 f/\partial x\, \partial y$ evaluated at the vertices as nodes (Fig. 7.12). This is a disadvantage in some cases. There are many problems in which first-derivative nodes can be reasonably estimated when they are not given explicitly, but the estimation of second-

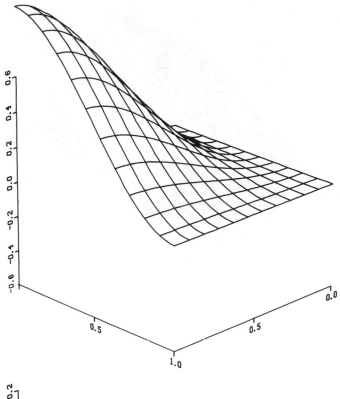

(a)

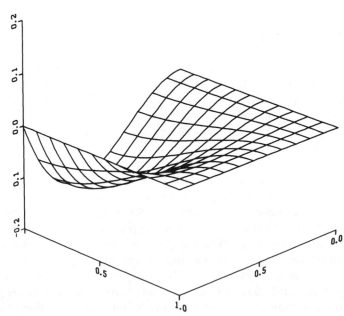

(b)

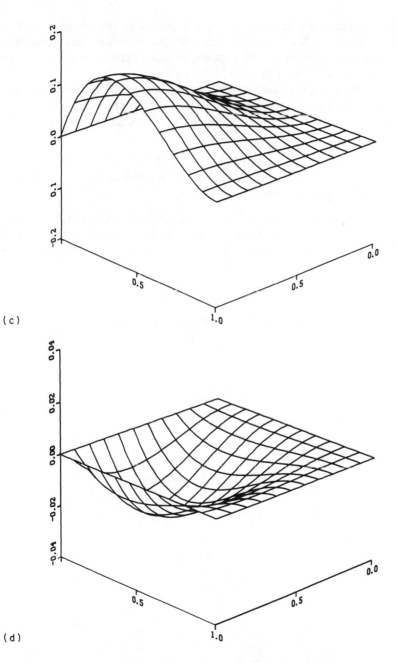

(c)

(d)

FIG. 9.6   Cardinal functions for the bicubic rectangle. (a) Node at $(1,0;0,0)$, (b) node at $(1,0;1,0)$, (c) node at $(1,0;0,1)$ and (d) node at $(1,0;1,1)$.

order derivatives introduces greater uncertainties to the procedure. We introduce a scheme in Section 9.9, which is $\mathscr{C}^1$, that avoids the use of second-order derivative nodes but at the expense of some further complication in the definition of the basis functions.

Let us summarize the details needed to implement the bicubic element. In this case it is convenient to use $[0, 1] \times [0, 1]$ as the standard geometrical figure. Thus, the standard finite element consists of the following:

(1)   the square $[0, 1] \times [0, 1]$;
(2)   the 16 nodes indicated in Fig. 7.12; and
(3)   the span of the 16 monomial functions listed in Fig. 7.12 (the "bicubic" polynomials).

We use Theorem 8.4.2 to generate the cardinal functions for interpolation on the standard element. First Eq. (2.9) applied to Hermite interpolation on $[0, 1]$ gives the following cardinal functions (again, we use the notation of Section 8.4 to specify the nodes):

Node	Cardinal function
$(0; 0)$	$(2\xi + 1)(\xi - 1)^2$
$(0; 1)$	$\xi(\xi - 1)^2$
$(1; 0)$	$\xi^2(3 - 2\xi)$
$(1; 1)$	$\xi^2(\xi - 1)$

TABLE 9.1
Cardinal Functions for the Bicubic Rectangle

Node	Cardinal function
$(0, 0; 0, 0)$	$(2\xi + 1)(\xi - 1)^2(2\eta + 1)(\eta - 1)^2$
$(0, 0; 1, 0)$	$\xi(\xi - 1)^2(2\eta + 1)(\eta - 1)^2$
$(0, 0; 0, 1)$	$(2\xi + 1)(\xi - 1)^2\eta\,(\eta - 1)^2$
$(0, 0; 1, 1)$	$\xi(\xi - 1)^2\eta(\eta - 1)^2$
$(1, 0; 0, 0)$	$\xi^2(3 - 2\xi)(2\eta + 1)(\eta - 1)^2$
$(1, 0; 1, 0)$	$\xi^2(\xi - 1)(2\eta + 1)(\eta - 1)^2$
$(1, 0; 0, 1)$	$\xi^2(3 - 2\xi)\eta(\eta - 1)^2$
$(1, 0; 1, 1)$	$\xi^2(\xi - 1)\eta(\eta - 1)^2$
$(1, 1; 0, 0)$	$\xi^2(3 - 2\xi)\eta^2(3 - 2\eta)$
$(1, 1; 1, 0)$	$\xi^2(\xi - 1)\eta^2(3 - 2\eta)$
$(1, 1; 0, 1)$	$\xi^2(3 - 2\xi)\eta^2(\eta - 1)$
$(1, 1; 1, 1)$	$\xi^2(\xi - 1)\eta^2(\eta - 1)$
$(0, 1; 0, 0)$	$(2\xi + 1)(\xi - 1)^2\eta^2(3 - 2\eta)$
$(0, 1; 1, 0)$	$\xi(\xi - 1)^2\eta^2(3 - 2\eta)$
$(0, 1; 0, 1)$	$(2\xi + 1)(\xi - 1)^2\eta^2(\eta - 1)$
$(0, 1; 1, 1)$	$\xi(\xi - 1)^2\eta^2(\eta - 1)$

Theorem 8.4.2 then gives the results of Table 9.1 for the standard bicubic rectangle. Four of these cardinal functions, all associated with the vertex $(1, 0)$, are illustrated in Fig. 9.6. The formulae of Table 9.1 may look formidable, but keep in mind that in numerical practice, these are stored once and for all in computer memory and evaluated as required.

Suppose, as in Section 9.2, that interpolation is to be carried out on $[x_0, x_M] \times [y_0, y_N]$ where $x_0 < x_1 < \ldots < x_M$ and $y_0 < y_1 < \ldots < y_N$, and the four items of data f, f_x, f_y, and f_{xy} are given at each of the $(M + 1)(N + 1)$ vertices of the array of sub-rectangles. (Here, subscripts denote partial derivatives.)

For a typical sub-rectangle $[x_j, x_{j+1}] \times [y_k, y_{k+1}]$ [call it the (j, k) sub-rectangle], there are then 16 associated nodes. To construct the interpolant on the (j, k) sub-rectangle we actually interpolate on the standard element, but we must be careful to adjust the nodal values involving derivatives to account for the change in size of the elements. The modifications required are indicated in Table 9.2. An interpolant is then formed on $[0, 1] \times [0, 1]$ using the modified nodal values and the cardinal functions of Table 9.1. The substitutions

$$\xi = (x - x_j)/(x_{j+1} - x_j), \qquad \eta = (y - y_k)/(y_{k+1} - y_k), \qquad (9.8)$$

then produce the bicubic interpolant on the (j, k) sub-rectangle. Of course, Eq. (9.8) defines the affine transformation mapping the (j, k) sub-rectangle onto $[0, 1] \times [0, 1]$.

Let us see how the computations proceed with rectangles 1 and 2 of Fig. 9.4, together with the function of Eq. (9.6) and Fig. 9.3. For rectangle 1, the Eq. (9.8) is

$$\xi = x - 1, \qquad \eta = 2y - 2. \qquad (9.9)$$

In Table 9.3 the nodal values for the surface of Fig. 9.3 on rectangle 1 are computed and listed in the first column. The third column shows the corresponding nodes of the standard element, and the second column gives the modified nodal values calculated according to Table 9.2.

TABLE 9.2
Modified Nodal Values

Nodal values for the (j, k) subrectangle	Nodal values for the standard element
f	f
f_x	$(x_{j+1} - x_j)f_x$
f_y	$(y_{k+1} - y_k)f_y$
f_{xy}	$(x_{j+1} - x_j)(y_{k+1} - y_k)f_{xy}$

TABLE 9.3
Modified Nodal Values for a Test Problem

Nodal values of rectangle (1)	Modified nodal values	Node
0.0	0.0	(0, 0; 0, 0)
−11.25	−11.25	(0, 0; 1, 0)
53.981	26.99	(0, 0; 0, 1)
−23.981	−11.99	(0, 0; 1, 1)
−11.25	−11.25	(1, 0; 0, 0)
−11.25	−11.25	(1, 0; 1, 0)
30.0	15.0	(1, 0; 0, 1)
30.0	15.0	(1, 0; 1, 1)
17.183	17.183	(0, 1; 0, 0)
−17.183	−17.183	(0, 1; 1, 0)
0.0007	0.00035	(0, 1; 0, 1)
15.0	7.5	(0, 1; 1, 1)
0.0	0.0	(1, 1; 0, 0)
0.0	0.0	(1, 1; 1, 0)
15.0	7.5	(1, 1; 0, 1)
15.0	7.5	(1, 1; 1, 1)

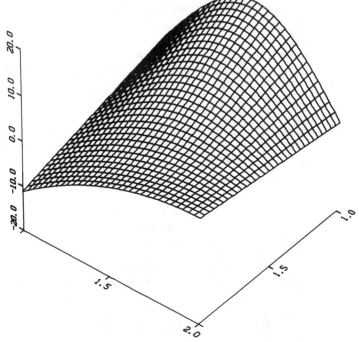

FIG. 9.7 Two bicubic rectangles and the model problem.

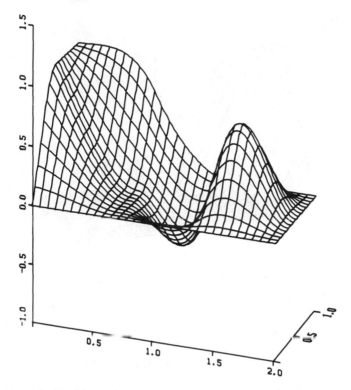

FIG. 9.8   Bicubic rectangle for the model problem: 3×2 sub-division.

To find the bicubic interpolant at a point x_0, y_0 of rectangle 1, first use Eq. (9.9) to compute $\xi_0 = x_0 - 1$ and $\eta_0 = 2y_0 - 2$, evaluate the cardinal functions of Table 9.1 at ξ_0, η_0 (these functions will, of course, generally be stored in computer memory), and multiply the resulting values by the corresponding nodal values of Table 9.3. Sum the resulting 16 products, and this gives $p(x_0, y_0)$. We repeat this for enough points on rectangle 1 to define the surface as closely as required.

We now repeat the whole process on rectangle 2. The result is illustrated in Fig. 9.7. The surface appears to be $\mathscr{C}^1$ (in contrast to Fig. 9.5) and seems to reproduce the model surface of Fig. 9.3 very well.

As a second, more ambitious, illustration we work with the model problem of Section 7.7. In Fig. 9.8 we indicate the result of applying this technique on a uniform sub-division of the base rectangle into $3 \times 2 = 6$ sub-rectangles. This implies a total of 48 nodal values, all of which are read from the analytically defined model problem. [Note that, even though $f(x, y)$ is not differentiable at $(0, 0)$, there are natural choices for the values

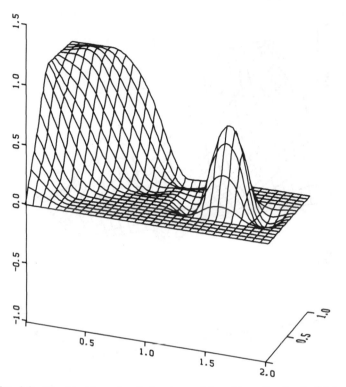

FIG. 9.9   The bicubic rectangle for the model problem: 6 × 4 sub-division.

of the first-derivative nodes obtained using "one-sided" derivatives. Also, the natural choice for f_{xy} here is zero.]

The effect of refining the lattice to $6 \times 4 = 24$ sub-rectangles is indicated in Fig. 9.9. This implies the use of 140 nodal values, which are, again, known precisely from the defining properties of the model surface. Other examples using the bicubic rectangle can be found in Section 9.11 and in Chapter 10.

9.4 The linear triangle

A scheme for linear interpolation on a triangle and its implementation with the use of a standard triangle has been discussed in some detail in Section 7.4. This procedure has great flexibility when viewed as the fundamental building block of a finite element scheme. To construct an interpolating surface to function values lying at the vertices of any polygonal domain and at a set of points inside this polygon, one may begin by triangulating the polygon in such a way that a function value is assigned at a point P if and only if P is a vertex of the triangulation. Then, we complete linear interpolation in each triangle (Theorem 7.4.2), and it is clear that a $\mathscr{C}^0$ surface consisting of plane triangular faces is formed.

If there are only four data points and they are at the vertices of a quadrilateral, it is already clear that there is more than one possible triangulation and that the surface generated depends on the choice of triangulation. We postpone discussion of the triangulation itself until Section 9.10.

A standard finite element for this scheme can be defined as consisting of

(1)   the equilateral triangle of Fig. 7.7,
(2)   the nodes $(-1, -\sqrt{3}/3; 0, 0)$, $(1, -\sqrt{3}/3; 0, 0)$, and $(0, 2\sqrt{3}; 0, 0)$, and
(3)   the class $\mathscr{P}_1$ of linear polynomials in ξ and η.

Then, given any triangle of the triangulation (take triangle T of Fig. 7.7 as typical), the function values assigned to the vertices of T are used to interpolate linearly on the standard triangle and to determine a function, say, $\hat{p}_T(\xi, \eta) = a + b\xi + c\eta$. The interpolant on T itself is then determined by substituting for ξ and η from Eq. (7.9).

As with the bilinear rectangle, the $\mathscr{C}^0$ smoothness of surfaces generated in this way may be acceptable if the data points are very dense (when triangulation is likely to be the major problem), but there are also many cases when $\mathscr{C}^0$ smoothness is not acceptable. Achieving $\mathscr{C}^1$ smooth surfaces with triangular finite elements requires more ingenuity of the interpolation scheme.

To illustrate, we consider interpolation to the surface of Fig. 9.3 whose domain $[1, 2] \times [1, 2]$ (cf. Fig. 9.4) is triangulated as in Fig. 9.10. First, use

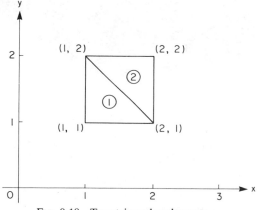

FIG. 9.10   Two triangular elements.

the technique described in Section 7.4 to map triangle 1 onto the standard triangle. We have to find the coefficients of Eq. (7.9). First, we let $(x_0, y_0) = (1, 1)$, $(x_1, y_1) = (2, 1)$, and $(x_2, y_2) = (1, 2)$. Then,

$$\Delta = \det \begin{bmatrix} 1 & 1 & 1 \\ 1 & 2 & 1 \\ 1 & 1 & 2 \end{bmatrix} = 1,$$

and it is found that

$$a_{11} = 2, \qquad a_{12} = 1, \qquad b_1 = -4,$$
$$a_{21} = 0, \qquad a_{22} = \sqrt{3}, \qquad b_2 = -4/3\sqrt{3}.$$

Thus, the required transformation is

$$\xi = 2x + y - 4, \qquad \eta = \sqrt{3}(y - 4/3). \tag{9.10}$$

The cardinal functions for interpolating on the standard triangle of Fig. 7.7 are easily found to be

$$\varphi_A(\xi, \eta) = \tfrac{1}{6}(-3\xi - \sqrt{3}\eta + 2),$$
$$\varphi_B(\xi, \eta) = \tfrac{1}{6}(3\xi - \sqrt{3}\eta + 2), \tag{9.11}$$
$$\varphi_C(\xi, \eta) = \tfrac{1}{3}(\sqrt{3}\eta + 1).$$

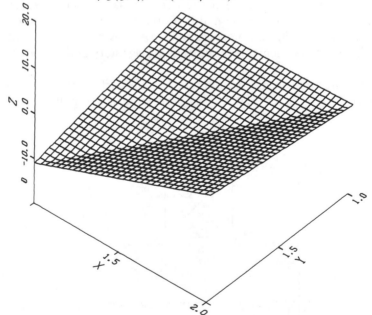

FIG. 9.11   Two linear triangles.

The nodal values at A, B, and C are obtained from Eq. (9.6):

$$f_A = 0.0, \qquad f_B = -11.25, \qquad f_C = 0.0005,$$

and the interpolant on the standard triangle is therefore

$$\hat{p}(\xi, \eta) = f_A \varphi_A(\xi, \eta) + f_B \varphi_B(\xi, \eta) + f_C \varphi_C(\xi, \eta).$$

The interpolant at points of triangle 1 is simply,

$$p(x, y) = \hat{p}(\xi, \eta),$$

as long as (x, y) and (ξ, η) are related as in Eq. (9.10). The result of this procedure applied to triangles 1 and 2 is illustrated in Fig. 9.11. It will be noted that the surface has "caved in".

9.5 Some higher degree triangular elements

In this section we make some brief comments on the schemes described in Section 7.5 as candidates for finite element techniques. We consider first the interpolation scheme described in the first paragraphs of Section 7.5. There are six function-value nodes, one at each vertex and one at the midpoint of each side. Interpolation at these nodes will reproduce all functions in $\mathscr{P}_2$ (the quadratic polynomials in two variables). The array of nodes for a triangulated polygon is illustrated in Fig. 9.12.

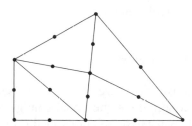

FIG. 9.12   Nodal array for quadratic triangles.

This scheme can certainly be used in a finite element procedure. Comparing this with the linear triangle scheme we see that there is, in general, a curved surface associated with each triangle. Each side common to two triangles supports a quadratic arc common to the two surfaces. However, the normal derivatives across the side need not be continuous. Also, at an interior vertex all directional derivatives may be discontinuous. Thus, the scheme generates a surface over the whole polygon, which is $\mathscr{C}^0$ but not generally $\mathscr{C}^1$.

The same is true of the more complicated schemes whose nodal configurations are indicated in Fig. 7.8 and which determine all polynomials in $\mathcal{P}_3$. The scheme of Fig. 7.8b will produce continuous directional derivatives at interior vertices, but otherwise, the directional derivative normal to interior edges need not be continuous.

It turns out that if we pursue this line of attack we must go as far as the "complete quintic" scheme of Fig. 7.9 before a $\mathcal{C}^1$ interpolating surface on the whole domain can be guaranteed. However, this requires complete information on the function values and first and second derivatives at all vertices, and this is rarely accessible in practice. Consequently, we simply note the feasibility of producing $\mathcal{C}^1$ surfaces using the complete quintic element; we do not investigate it in any more detail but go on to other schemes that do not require values of second derivatives as nodal values.

To accomplish this, it is necessary to change the basis functions of a finite element in some way, since the geometrical figure and the nodal configuration are essentially prescribed. We shall discuss one way of modifying the basis functions: that of introducing "seams" into the geometrical figure. Another way is to admit some rational functions (quotients of polynomials) as well as polynomials into the class of functions which can be reproduced by interpolation. One approach due to Zienciewicz and to Mitchell and Wait is described by Mitchell and Wait (1977). Other lines of attack developed by Birkhoff, Mansfield, Barnhill, Gordon, and others, using rational functions and blending techniques are described by Barnhill (1977), as are several other approaches. Alfeld and Barnhill (1984) give a construction for a $\mathcal{C}^2$ triangular element.

9.6 The biquadratic seamed element

For the purpose of illustration, we describe a "seamed" rectangular element, which has precisely the same nodal configuration as the bicubic rectangle but different associated functions in the interpolation scheme. We consider first an interpolation problem posed on an interval $[x_0, x_1]$. The nodes concerned are simply the four nodes of Hermitian interpolation: $(x_0; 0)$, $(x_0; 1)$, $(x_1; 0)$, and $(x_1; 1)$. We know that in conjunction with the class $\mathcal{P}_3$ of cubic polynomials there is a well-defined interpolation scheme. We retain these nodes but change the admissible functions to the class $\mathcal{Q}_2$ defined in the following way: The function $q(x) \in \mathcal{Q}_2$ if and only if it is a quadratic polynomial on $[x_0, \frac{1}{2}(x_0 + x_1)]$ and a (generally different) quadratic polynomial on $[\frac{1}{2}(x_0 + x_1), x_1]$ and $q \in \mathcal{C}^1[a, b]$. Thus, there are constants $a_1, b_1, c_1, a_2, b_2,$ and c_2 such that

$$q(x) = \begin{cases} a_1 + b_1 x + c_1 x^2 & \text{if } x \in [x_0, \frac{1}{2}(x_0 + x_1)], \\ a_2 + b_2 x + c_2 x^2 & \text{if } x \in [\frac{1}{2}(x_0 + x_1), x_1], \end{cases}$$

and these constants are chosen so that $q(x)$ and the derivative $q'(x)$ are continuous at the mid-point of $[x_0, x_1]$. In general, $q(x) \notin \mathscr{C}^2[x_0, x_1]$. These constraints can be shown to imply that, like $\mathscr{P}_3$, the class $\mathscr{Q}_2$ has dimension four. Furthermore, it is not hard to verify that the interpolation problem posed on $[x_0, x_1]$ is well defined: Given nodal values $f_0, f_0', f_1,$ and f_1' at the respective nodes listed above, there is a unique $q(x) \in \mathscr{Q}_2$ such that

$$q(x_0) = f_0, \qquad q'(x_0) = f_0', \qquad q(x_1) = f_1, \qquad q'(x_1) = f_1'.$$

Now recall Example 3 of Section 8.5. There, the classical Hermitian scheme on an interval is employed with the tensor product process to obtain the bicubic rectangle. In just the same way, our scheme using the functions of $\mathscr{Q}_2$ generates an interpolation process on a rectangle $[x_0, x_1] \times [y_0, y_1]$. Indeed, the nodal configuration is precisely that of the bicubic rectangle and is indicated in Fig. 7.12. The basis functions are now *biquadratic* polynomials however, one in each quarter of the rectangle $[x_0, x_1] \times [y_0, y_1]$ (the functions $r_1, r_2, r_3,$ and r_4 of Fig. 9.13). All the basis functions and their first partial derivatives are continuous on the *complete* rectangle.

The two bisectors here are sometimes referred to as the seams of the element. Thus, any interpolating function defined by this element is $\mathscr{C}^1$ on the whole element and may have discontinuities in the *second* derivatives on the seams.

This element could now be used in just the same way as the bicubic rectangle to produce $\mathscr{C}^1$ interpolating surfaces over a mesh of adjoining rectangles, given the data f, f_x, f_y and f_{xy} at each vertex of the mesh. Our main purpose in discussing this element has been to introduce the idea of a seamed element and not to supply the reader with a widely useful technique. We therefore present no further details and go on to discussion of rather more complicated but useful elements.

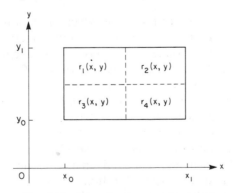

Fig. 9.13 Biquadratic seamed rectangle.

9.7 The Clough–Tocher triangle

The element to be described here was developed by R. W. Clough and J. L. Tocher (1965) for use in problems in structural mechanics. Some of its theoretical properties were established by P. Percell (1976), and our exposition is based on the work of S. Ritchie (1978).

As in Sections 9.4 and 9.5, we suppose that an interpolation problem is posed on a polygonal domain, and that this domain has been triangulated. We are to describe an interpolation scheme which, when applied on each triangle, will generate a $\mathscr{C}^1$ surface over the complete polygonal domain. It has been remarked in Section 9.5 that this can always be done by means of the complete quintic scheme described in Section 7.5. However, this has 21 nodes for each triangle (many of them shared with other triangles, of course) and requires second-derivative data at all vertices. By using a seaming technique, the Clough–Tocher triangle achieves $\mathscr{C}^1$ interelement continuity with only 12 nodes for each triangle and with cubic polynomials.

The Clough–Tocher triangle has seams that are lines joining each vertex to the centroid of the triangle, thus dividing the triangular element into three subtriangles (Fig. 9.14). The functions admitted in the interpolation

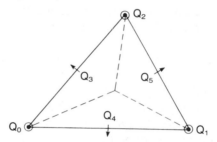

FIG. 9.14   Nodal configuration for the Clough–Tocher triangle.

belong to the class $\mathscr{H}_3$, say, of functions which are (bivariate) cubic polynomials in each sub-triangle *and* are continuous, together with their first derivatives over the whole triangle. The class $\mathscr{H}_3$ also depends on the choice of triangle, of course. Thus, a function in $\mathscr{H}_3$ is of class $\mathscr{C}^1$ over the triangular element. It turns out that the dimension of this class $\mathscr{H}_3$ is 12.

Now we introduce the 12 nodes indicated in Fig. 9.14. The nodal values are, say, f, f_x, and f_y at each vertex and a derivative $\partial f/\partial n$ at the mid-point of each side in the direction of the normal to the side. It is a remarkable fact that the 12 freedoms of $\mathscr{H}_3$ match these 12 nodes to produce a well-defined interpolation scheme. Thus, given the 12 nodal values on any triangle, there is a unique interpolating function, which is $\mathscr{C}^1$ on the whole triangle, and a cubic polynomial in each sub-triangle.

To use the Clough–Tocher triangle to generate a $\mathscr{C}^1$ surface over a triangulated region, it is also necessary that, when applied on adjacent triangles, a $\mathscr{C}^1$ surface be generated on the union of the triangles. Again, it turns out that this is the case.

To implement the technique, it remains to consider the problem of transforming the interpolation on each triangle to interpolation on a standard triangle and to indicate how this is done in terms of the 12 associated cardinal functions.

We use the standard triangle of Fig. 7.7, so that the mapping of an arbitrary triangle onto this one is defined by Eq. (7.9). Note that a function of $\mathscr{K}_3$ will, under this transformation, correspond to a $\mathscr{C}^1$ function on the standard triangle, which is again a cubic polynomial on each sub-triangle. (This is sometimes described as an "affine invariance" property.)

To list the cardinal functions for interpolation on the standard triangle, the 12 nodes must be separately identified. This is done by means of Fig. 9.15. Note that the first-derivative nodes at the vertices are specified in the direction of the radius from the centroid to the vertex and the (clockwise) orthogonal direction.

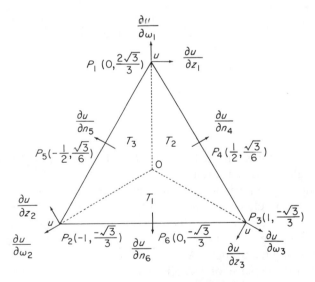

FIG. 9.15   Clough–Tocher standard element.

Let T and T_0 denote an arbitrary triangle and the standard triangle, respectively. Suppose that $f(x, y)$ is the function in $\mathscr{K}_3(T)$ to be found and $u(\xi, \eta)$ is the corresponding function in $\mathscr{K}_3(T_0)$. Thus, u is determined from f by substituting the expressions in Eq. (7.10) for x and y in $f(x, y)$. Let us first consider how the nodes are transformed by supposing that values of f and the partial derivatives f_x and f_y are given at each vertex of T and (although there is some redundancy here) that f_x and f_y are given at the mid-points of the sides of T. The corresponding values of the derivative nodes on T_0 (indicated in Fig. 9.15) are obtained in two steps. First, the chain rule implies that, at any point of T_0,

$$u_\xi = b_{11}f_x + b_{21}f_y, \qquad u_\eta = b_{12}f_x + b_{22}f_y, \tag{9.12}$$

and b_{11}, b_{21}, b_{12}, and b_{22} are determined by the vertices of T as in Eq. (7.10). Second, the required nodal values are determined in terms of the dimensions of the original triangle (see Fig. 9.14), the given mid-point normal derivatives and values of u_ξ and u_η by Eq. (9.13):

$$\frac{\partial u}{\partial z_1} = u_\xi(P_1), \qquad \frac{\partial u}{\partial z_2} = \left[-\frac{1}{2}u_\xi + \frac{\sqrt{3}}{2}u_\eta\right](P_2), \qquad \frac{\partial u}{\partial z_3} = \left[-\frac{1}{2}u_\xi - \frac{\sqrt{3}}{2}u_\eta\right](P_3)$$

$$\frac{\partial u}{\partial w_1} = u_\eta(P_1), \qquad \frac{\partial u}{\partial w_2} = \left[-\frac{\sqrt{3}}{2}u_\xi - \frac{1}{2}u_\eta\right](P_2), \qquad \frac{\partial u}{\partial w_3} = \left[\frac{\sqrt{3}}{2}u_\xi - \frac{1}{2}u_\eta\right](P_3)$$

$$\frac{\partial u}{\partial n_4} = \frac{d_{02}^2 - d_{01}^2}{\sqrt{3}d_{12}^2}\left\{\frac{3}{4}[u(P_1) - u(P_3)] + \frac{1}{8}\left[\frac{\partial u}{\partial z_1} + \frac{\partial u}{\partial z_3}\right] - \frac{\sqrt{3}}{8}\left[\frac{\partial u}{\partial w_1} - \frac{\partial u}{\partial w_3}\right]\right\} + \frac{\Delta}{\sqrt{3}d_{12}}\frac{\partial f}{\partial n}(Q_5),$$

$$\frac{\partial u}{\partial n_5} = \frac{d_{01}^2 - d_{12}^2}{\sqrt{3}d_{02}^2}\left\{\frac{3}{4}[u(P_2) - u(P_1)] + \frac{1}{8}\left[\frac{\partial u}{\partial z_2} + \frac{\partial u}{\partial z_1}\right] - \frac{\sqrt{3}}{8}\left[\frac{\partial u}{\partial w_2} - \frac{\partial u}{\partial w_1}\right]\right\} + \frac{\Delta}{\sqrt{3}d_{02}}\frac{\partial f}{\partial n}(Q_3),$$

$$\frac{\partial u}{\partial n_6} = \frac{d_{12}^2 - d_{02}^2}{\sqrt{3}d_{01}^2}\left\{\frac{3}{4}[u(P_3) - u(P_2)] + \frac{1}{8}\left[\frac{\partial u}{\partial z_2} + \frac{\partial u}{\partial z_3}\right] - \frac{\sqrt{3}}{8}\left[\frac{\partial u}{\partial w_3} - \frac{\partial u}{\partial w_2}\right]\right\} + \frac{\Delta}{\sqrt{3}d_{01}}\frac{\partial f}{\partial n}(Q_4).$$

$$\tag{9.13}$$

The symbol d_{ij} represents the length of the side connecting Q_i to Q_j on the original triangle and Δ is its area.

The Clough–Tocher triangle is sometimes simplified to what is, effectively, a *nine*-node element. These nodes are the three required at each vertex. The normal derivative at the mid-side is then estimated as the mean of the corresponding normal derivatives at the two vertices associated with the side on the original triangle. A considerable amount of labour is saved in this way but at the expense of a smaller class of polynomials reproduced by the interpolant. The full Clough–Tocher triangle will reproduce all cubic polynomials, whereas, with this approximation, all quadratic polynomials are reproduced and some cubic polynomials are not.

We now know how to transfer the nodal values from T to T_0 and, in combination with the 12 corresponding cardinal functions on T_0 [which form a basis for $\mathcal{H}_3(T_0)$], this allows us to determine $u(\xi, \eta)$ completely.

The interpolant $f(x, y)$ on T is then obtained by substituting from Eq. (7.9) for ξ and η in u. It only remains for us to list the twelve cardinal functions. Their values are, of course, zero at all the nodes except one. In the following list, this exceptional node is indicated in the right-hand column. Triangles T_1, T_2 and T_3 are the sub-triangles determined by the seams, as shown in Fig. 9.15:

$$B_1(\xi, \eta) = \frac{1}{3} + \frac{\sqrt{3}}{2}\eta$$

$$+ \begin{cases} -\dfrac{\sqrt{3}}{2}\eta^3, & \text{in } T_1; \\[2ex] -\dfrac{1}{4}\xi^3 - \dfrac{3\sqrt{3}}{8}\xi^2\eta - \dfrac{\sqrt{3}}{8}\eta^3, & \text{in } T_2, \quad B_1(P_1) = 1; \\[2ex] \dfrac{1}{4}\xi^3 - \dfrac{3\sqrt{3}}{8}\xi^2\eta - \dfrac{\sqrt{3}}{8}\eta^3, & \text{in } T_3. \end{cases}$$

$$B_2(\xi, \eta) = \frac{1}{3} - \frac{3}{4}\xi - \frac{\sqrt{3}}{4}\eta$$

$$+ \begin{cases} \dfrac{1}{4}\xi^3 + \dfrac{\sqrt{3}}{4}\eta^3, & \text{in } T_1; \\[2ex] \dfrac{9}{16}\xi^3 + \dfrac{9\sqrt{3}}{16}\xi^2\eta + \dfrac{9}{16}\xi\eta^2 + \dfrac{\sqrt{3}}{16}\eta^3 & \text{in } T_2, \quad B_2(P_2) = 1; \\[2ex] \dfrac{5}{16}\xi^3 - \dfrac{3\sqrt{3}}{16}\xi^2\eta + \dfrac{9}{16}\xi\eta^2 + \dfrac{\sqrt{3}}{16}\eta^3 & \text{in } T_3. \end{cases}$$

$$B_3(\xi, \eta) = \frac{1}{3} + \frac{3}{4}\xi - \frac{\sqrt{3}}{4}\eta$$

$$+ \begin{cases} -\dfrac{1}{4}\xi^3 + \dfrac{\sqrt{3}}{4}\eta^3, & \text{in } T_1; \\[2ex] -\dfrac{5}{16}\xi^3 - \dfrac{3\sqrt{3}}{16}\xi^2\eta - \dfrac{9}{16}\xi\eta^2 + \dfrac{\sqrt{3}}{16}\eta^3 & \text{in } T_2, \quad B_3(P_3) = 1; \\[2ex] -\dfrac{9}{16}\xi^3 + \dfrac{9\sqrt{3}}{16}\xi^2\eta - \dfrac{9}{16}\xi\eta^2 + \dfrac{\sqrt{3}}{16}\eta^3 & \text{in } T_3. \end{cases}$$

$$B_4(\xi, \eta) = \frac{1}{4}\xi + \frac{\sqrt{3}}{2}\xi\eta$$

$$+\begin{cases} \dfrac{3}{4}\xi\eta^2, & \text{in } T_1; \\[2mm] -\dfrac{5}{16}\xi^3 - \dfrac{5\sqrt{3}}{8}\xi^2\eta - \dfrac{3}{16}\xi\eta^2, & \text{in } T_2, \quad \dfrac{\partial B_4}{\partial z_1}(P_1) = 1; \\[2mm] -\dfrac{5}{16}\xi^3 + \dfrac{5\sqrt{3}}{8}\xi^2\eta - \dfrac{3}{16}\xi\eta^2, & \text{in } T_3. \end{cases}$$

$$B_5(\xi, \eta) = -\frac{1}{8}\xi + \frac{\sqrt{3}}{8}\eta + \frac{3}{8}\xi^2 - \frac{\sqrt{3}}{4}\xi\eta - \frac{3}{8}\eta^2$$

$$+\begin{cases} -\dfrac{1}{8}\xi^3 + \dfrac{\sqrt{3}}{4}\xi^2\eta - \dfrac{3\sqrt{3}}{8}\eta^3, & \text{in } T_1; \\[2mm] -\dfrac{9}{32}\xi^3 + \dfrac{3\sqrt{3}}{32}\xi^2\eta + \dfrac{15}{32}\xi\eta^2 + \dfrac{3\sqrt{3}}{32}\eta^3, & \text{in } T_2, \quad \dfrac{\partial B_5}{\partial z_2}(P_2) = 1; \\[2mm] \dfrac{11}{32}\xi^3 - \dfrac{17\sqrt{3}}{32}\xi^2\eta + \dfrac{15}{32}\xi\eta^2 + \dfrac{3\sqrt{3}}{32}\eta^3, & \text{in } T_3. \end{cases}$$

$$B_6(\xi, \eta) = -\frac{1}{8}\xi - \frac{\sqrt{3}}{8}\eta - \frac{3}{8}\xi^2 - \frac{\sqrt{3}}{4}\xi\eta + \frac{3}{8}\eta^2$$

$$+\begin{cases} -\dfrac{1}{8}\xi^3 - \dfrac{\sqrt{3}}{4}\xi^2\eta + \dfrac{3\sqrt{3}}{8}\eta^3, & \text{in } T_1; \\[2mm] \dfrac{11}{32}\xi^3 + \dfrac{17\sqrt{3}}{32}\xi^2\eta + \dfrac{15}{32}\xi\eta^2 - \dfrac{3\sqrt{3}}{32}\eta^3, & \text{in } T_2, \quad \dfrac{\partial B_6}{\partial z_3}(P_3) = 1; \\[2mm] -\dfrac{9}{32}\xi^3 - \dfrac{3\sqrt{3}}{32}\xi^2\eta + \dfrac{15}{32}\xi\eta^2 - \dfrac{3\sqrt{3}}{32}\eta^3, & \text{in } T_3. \end{cases}$$

$$B_7(\xi, \eta) = -\frac{7\sqrt{3}}{81} - \frac{13}{36}\eta + \frac{\sqrt{3}}{18}\eta^2$$

$$+\begin{cases} \dfrac{17}{36}\eta^3, & \text{in } T_1; \\[2mm] \dfrac{41}{144}\eta^3 + \dfrac{1}{8\sqrt{3}}\xi^3 + \dfrac{3}{16}\xi^2\eta, & \text{in } T_2, \quad \dfrac{\partial B_7}{\partial w_1}(P_1) = 1; \\[2mm] \dfrac{41}{144}\eta^3 - \dfrac{1}{8\sqrt{3}}\xi^3 + \dfrac{3}{16}\xi^2\eta, & \text{in } T_3. \end{cases}$$

$$B_8(\xi, \eta) = -\frac{7\sqrt{3}}{81} + \frac{13\sqrt{3}}{72}\xi + \frac{13}{72}\eta + \frac{\sqrt{3}}{24}\xi^2 + \frac{1}{12}\xi\eta + \frac{\sqrt{3}}{72}\eta^2$$

$$+ \begin{cases} -\dfrac{\sqrt{3}}{8}\xi^3 - \dfrac{1}{4}\xi^2\eta - \dfrac{\sqrt{3}}{12}\xi\eta^2 - \dfrac{11}{72}\eta^3, & \text{in } T_1; \\[2ex] -\dfrac{17\sqrt{3}}{96}\xi^3 - \dfrac{17}{32}\xi^2\eta - \dfrac{17\sqrt{3}}{96}\xi\eta^2 - \dfrac{17}{288}\eta^3, & \text{in } T_2, \quad \dfrac{\partial B_8}{\partial w_2}(P_2) = 1; \\[2ex] -\dfrac{13\sqrt{3}}{96}\xi^3 - \dfrac{5}{32}\xi^2\eta - \dfrac{17\sqrt{3}}{96}\xi\eta^2 - \dfrac{17}{288}\eta^3, & \text{in } T_3. \end{cases}$$

$$B_9(\xi, \eta) = -\frac{7\sqrt{3}}{81} - \frac{13\sqrt{3}}{72}\xi + \frac{13}{72}\eta + \frac{\sqrt{3}}{24}\xi^2 + \frac{\sqrt{3}}{72}\eta^2 - \frac{1}{12}\xi\eta$$

$$+ \begin{cases} \dfrac{3\sqrt{3}}{24}\xi^3 - \dfrac{1}{4}\xi^2\eta + \dfrac{\sqrt{3}}{12}\xi\eta^2 - \dfrac{11}{72}\eta^3, & \text{in } T_1; \\[2ex] \dfrac{13\sqrt{3}}{96}\xi^3 - \dfrac{5}{32}\xi^2\eta + \dfrac{17\sqrt{3}}{96}\xi\eta^2 - \dfrac{17}{288}\eta^3, & \text{in } T_2, \quad \dfrac{\partial B_9}{\partial w_3}(P_3) = 1; \\[2ex] \dfrac{17\sqrt{3}}{96}\xi^3 - \dfrac{17}{32}\xi^2\eta + \dfrac{17\sqrt{3}}{96}\xi\eta^2 - \dfrac{17}{288}\eta^3, & \text{in } T_3 \end{cases}$$

$$B_{10}(\xi, \eta) = -\frac{4\sqrt{3}}{81} - \frac{\sqrt{3}}{9}\xi - \frac{1}{9}\eta - \frac{2}{3}\xi\eta + \frac{2\sqrt{3}}{9}\eta^2$$

$$+ \begin{cases} \dfrac{5}{9}\eta^3 - \dfrac{\sqrt{3}}{3}\xi\eta^2, & \text{in } T_1; \\[2ex] -\dfrac{7}{36}\eta^3 + \dfrac{5\sqrt{3}}{12}\xi^3 + \dfrac{9}{4}\xi^2\eta + \dfrac{5\sqrt{3}}{12}\xi\eta^2, & \text{in } T_2, \quad \dfrac{\partial B_{10}}{\partial n_4}(P_4) = 1; \\[2ex] -\dfrac{7}{36}\eta^3 + \dfrac{\sqrt{3}}{12}\xi^3 - \dfrac{3}{4}\xi^2\eta + \dfrac{5\sqrt{3}}{12}\xi\eta^2, & \text{in } T_3. \end{cases}$$

$$B_{11}(\xi, \eta) = -\frac{4\sqrt{3}}{81} + \frac{\sqrt{3}}{9}\xi - \frac{1}{9}\eta + \frac{2\sqrt{3}}{9}\eta^2 + \frac{2}{3}\xi\eta$$

$$+ \begin{cases} \dfrac{5}{9}\eta^3 + \dfrac{\sqrt{3}}{3}\xi\eta^2, & \text{in } T_1; \\[2ex] -\dfrac{\sqrt{3}}{12}\xi^3 - \dfrac{7}{36}\eta^3 - \dfrac{3}{4}\xi^2\eta - \dfrac{5\sqrt{3}}{12}\xi\eta^2, & \text{in } T_2, \dfrac{\partial B_{11}}{\partial n_5}(P_5) = 1; \\[2ex] -\dfrac{5\sqrt{3}}{12}\xi^3 - \dfrac{7}{36}\eta^3 + \dfrac{9}{4}\xi^2\eta - \dfrac{5\sqrt{3}}{12}\xi\eta^2, & \text{in } T_3 \end{cases}$$

$$B_{12}(\xi, \eta) = -\frac{4\sqrt{3}}{81} + \frac{2}{9}\eta + \frac{\sqrt{3}}{3}\xi^2 - \frac{\sqrt{3}}{9}\eta^2$$

$$+ \begin{cases} \xi^2\eta - \dfrac{13}{9}\eta^3, & \text{in } T_1; \\[2mm] -\dfrac{\sqrt{3}}{3}\xi^3 - \dfrac{1}{2}\xi^2\eta + \dfrac{1}{18}\eta^3, & \text{in } T_2, \quad \dfrac{\partial B_{12}}{\partial n_6}(P_6) = 1; \\[2mm] \dfrac{\sqrt{3}}{3}\xi^3 - \dfrac{1}{2}\xi^2\eta + \dfrac{1}{18}\eta^3, & \text{in } T_3. \end{cases}$$

Let us illustrate the use of this finite element with data drawn from the surface of Fig. 9.3 and Eq. (9.6). The triangulation of Fig. 9.10 is adopted. Thus, to interpolate on triangle 1 of Fig. 9.10, we first need Eq. (9.10),

$$\xi = 2x + y - 4, \qquad \eta = \sqrt{3}(y - \tfrac{4}{3}), \tag{9.14}$$

which maps triangle 1 onto the standard equilateral triangle T_0 of Fig. 9.15.

The coefficients b_{11}, b_{12}, b_{21} and b_{22} of the inverse transformation [Eq. (7.10)] are

$$b_{11} = \tfrac{1}{2}, \qquad b_{12} = -1/2\sqrt{3}, \qquad b_{21} = 0, \qquad b_{22} = 1/\sqrt{3}.$$

These numbers are used in Eq. (9.12), which, together with Eq. (9.13), determines the transformed nodal data. Of course, in numerical practice this is automated, so that all calculations proceed in exactly the same way once the vertices of a triangle and the nodal values are specified.

In Tables 9.4 and 9.5 we show that the nodal values on triangle 1, determined from Eq. (9.6), and the transformed nodal values for triangle T_0, obtained by means of Eq. (9.12) and (9.13).

To evaluate the value of the interpolant $p(x, y)$ at any point (x_0, y_0) of triangle 1, we first find the corresponding point (ξ_0, η_0) of T_0 by using Eq. (9.14). Evaluate each of the 12 cardinal functions at (ξ_0, η_0), i.e. $B_j(\xi_0, \eta_0)$,

TABLE 9.4
Nodal Values on Triangle 1

Position	f	f_x	f_y
Vertex (1, 2)	0.0005	3.75	−53.981
Vertex (1, 1)	0.0	−11.25	53.981
Vertex (2, 1)	−11.25	−11.25	30.0
Mid-side ($\tfrac{3}{2}$, $\tfrac{3}{2}$)	6.487	−6.487	7.5
Mid-side (1, $\tfrac{3}{2}$)	17.183	−17.183	0.0007
Mid-side ($\tfrac{3}{2}$, 1)	−5.625	−11.25	35.38

TABLE 5
Transformed Nodal Values on Triangle T_0

Vertex P_1	$u = 0.0005$	$\dfrac{\partial u}{\partial z_1} = 1.875$		$\dfrac{\partial u}{\partial w_1} = -34.414$
Vertex P_2	$u = 0.0$	$\dfrac{\partial u}{\partial z_2} = 32.615$		$\dfrac{\partial u}{\partial w_2} = -12.335$
Vertex P_3	$u = -11.25$	$\dfrac{\partial u}{\partial z_3} = -26.991$		$\dfrac{\partial u}{\partial w_3} = -15.155$
Mid-side P_4			$\dfrac{\partial u}{\partial n_1} = 12.335$	
Mid-side P_5			$\dfrac{\partial u}{\partial n_2} = -15.155$	
Mid-side P_6			$\dfrac{\partial u}{\partial n_3} = -23.674$	

$j = 1, 2, \ldots, 12$. Multiply each function value by the corresponding nodal value of Table 9.5 and add the results to get $\hat{p}(\xi_0, \eta_0)$. Then, we have $p(x_0, y_0) = \hat{p}(\xi_0, \eta_0)$.

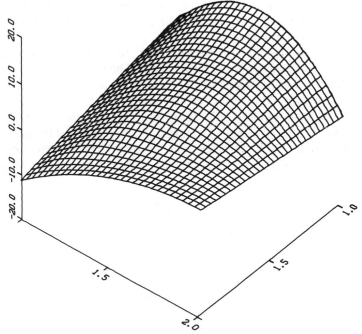

FIG. 9.16   Model problem represented by two Clough–Tocher triangles.

The result of applying this procedure to our model problem on triangles 1 and 2 is illustrated in Fig. 9.16. The result should be compared with the interpolated surface of Fig. 9.3, and the apparent $\mathscr{C}^1$ nature of the interpolated surface should be noted.

Further illustrations of the use of the Clough–Tocher triangle occur in Chapter 10, where it is used in combination with a moving least squares technique.

9.8 Piecewise quadratic triangular elements

Surface construction is often followed by the development of contour maps to represent the surface. For this purpose it is advantageous to have the surface composed of quadratic pieces, such as triangular patches (or finite elements), for example, on each of which the surface is represented by an associated quadratic function. At the same time the whole surface should be $\mathscr{C}^1$. It has been remarked in Section 9.5 that the quadratic triangle scheme does not produce a $\mathscr{C}^1$ surface over several triangles. Indeed, in order to produce $\mathscr{C}^1$ smoothness on triangles, we first noted the complete quintic, which is very complicated and has too many nodes, and in Section 9.7, we succeeded in producing $\mathscr{C}^1$ surfaces but with cubic patches.

There does not seem to be a satisfactory solution to the construction of a general-purpose finite element which is "patch-wise" quadratic and globally $\mathscr{C}^1$ [Powell and Sabin (1977)]. However, there are some special cases in which these two properties can be achieved. The most general is probably the case in which all triangles are known to be acute. We shall confine attention to two simple cases of *regular* triangulations.

Consider first a domain tessellated by isosceles triangles, as in Fig. 9.17. We associate with these isosceles triangles the equilateral standard triangle of Fig. 9.18 (and Fig. 7.7). Here, the dotted lines are the three perpendiculars from vertices to opposite faces, and they divide the triangle

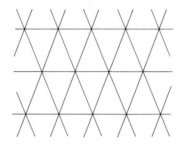

Fig. 9.17   Tessellation of isosceles triangles.

into six subtriangles. The admissible functions are those that are $\mathscr{C}^1$ on the whole triangle and quadratic in each patch. It turns out that the space of these functions has dimension nine, and matched with the nine nodes of interpolation indicated in Fig. 9.18, a well-defined interpolation procedure results. Furthermore, when applied to a *regular* tessellation like that of Fig. 9.17, a globally $\mathscr{C}^1$ interpolant is produced.

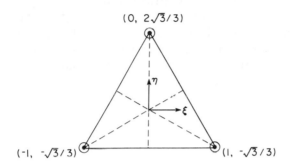

$(0, 2\sqrt{3}/3)$

η

ξ

$(-1, -\sqrt{3}/3)$ $(1, -\sqrt{3}/3)$

FIG. 9.18   Standard piecewise quadratic triangle T_0.

If T is a triangle of the tessellation, the coordinates of the vertices will either have the form

$$(x_0, y_0), \qquad (x_1, y_0), \qquad (\tfrac{1}{2}(x_0 + x_1), y_1), \qquad (9.15)$$

where $x_1 > x_0$ and $y_1 > y_0$, or the form

$$(\tfrac{1}{2}(x_0 + x_1), y_1), \qquad (\tfrac{1}{2}(3x_1 - x_0), y_1), \qquad (x_1, y_0). \qquad (9.16)$$

For a triangle with vertices of the form of Eq. (9.15), T is mapped onto T_0 by means of

$$\xi = 3\left(\frac{x - x_0}{x_1 - x_0}\right) - 1, \qquad \eta = \sqrt{3}\left(\frac{y - y_0}{y_1 - y_0} - \frac{1}{3}\right), \qquad (9.17)$$

and in case Eq. (9.16) applies:

$$\xi = \left(\frac{x_1 + x_0 - 2x}{x_1 - x_0}\right) + 1, \qquad \eta = \sqrt{3}\left(\frac{y_0 - y}{y_1 - y_0} + \frac{2}{3}\right). \qquad (9.18)$$

The nine given nodal values on T are now to be transferred to T_0 with scaling of the derivative nodes, as indicated by Eq. (9.17) or Eq. (9.18). As one might expect, the cardinal functions are quite difficult to describe. We do it in two stages [following Ritchie (1978)]. We first define nine functions $g_1, \ldots, g_9$ and then define the cardinal functions $B_1, \ldots, B_9$ in terms of them.

It will be convenient to introduce the truncated function denoted by a subscript $+$ as follows:

$$s_+ = \begin{cases} s & \text{if} \quad s \geq 0, \\ 0 & \text{if} \quad s < 0. \end{cases} \tag{9.19}$$

The function $(x + y)_+$ in the x-, y-coordinate plane then takes the value $x + y$ when $x + y \geq 0$, i.e. to the right of the line $y = -x$, and takes the identically zero value when $x + y < 0$, or to the left of the line $y = -x$.

We will need functions defined by expressions like ξ_+^2, or $(\xi - \eta)_+^2$. Here, it is to be understood that the operation of truncation, denoted by the subscript $+$, is performed first and is followed by the operation of exponentiation.

Now define,

$$g_1(\xi, \eta) = 1, \qquad\qquad\qquad g_2(\xi, \eta) = \xi,$$

$$g_3(\xi, \eta) = \eta, \qquad\qquad\qquad g_4(\xi, \eta) = \xi_+^2,$$

$$g_5(\xi, \eta) = (-\xi)_+^2, \qquad\qquad g_6(\xi, \eta) = \left(\eta - \frac{\sqrt{3}}{3}\xi\right)_+^2,$$

$$g_7(\xi, \eta) = \left(\frac{\sqrt{3}}{3}\xi - \eta\right)_+^2, \qquad g_8(\xi, \eta) = \left(\eta + \frac{\sqrt{3}}{3}\xi\right)_+^2,$$

$$g_9(\xi, \eta) = \left(-\eta - \frac{\sqrt{3}}{3}\xi\right)_+^2.$$

It is easily seen that each of these functions is a quadratic on each subtriangle of the standard triangle and is $\mathscr{C}^1$ on the whole triangle.

The nine cardinal functions can now be defined as follows:

$$B_1(\xi, \eta) = \frac{1}{3} + \frac{2\sqrt{3}}{3}\eta - \frac{1}{3}g_4 - \frac{1}{3}g_5 - \frac{1}{4}g_6 + \frac{1}{2}g_7 - \frac{1}{4}g_8 - \frac{1}{2}g_9,$$

$$B_1(P_1) = 1;$$

$$B_2(\xi, \eta) = \frac{1}{3} - \xi - \frac{\sqrt{3}}{3}\eta + \frac{2}{3}g_4 - \frac{1}{3}g_5 - \frac{1}{4}g_6 - \frac{1}{4}g_7 + \frac{1}{2}g_8 - \frac{1}{4}g_9,$$

$$B_2(P_2) = 1;$$

$$B_3(\xi, \eta) = \frac{1}{3} + \xi - \frac{\sqrt{3}}{3}\eta - \frac{1}{3}g_4 + \frac{2}{3}g_5 + \frac{1}{2}g_6 - \frac{1}{4}g_7 - \frac{1}{4}g_8 - \frac{1}{4}g_9,$$

$$B_3(P_3) = 1;$$

$$B_4(\xi, \eta) = -\frac{3}{8}g_6 + \frac{3}{8}g_8,$$

$$\partial B_4 / \partial \xi (P_1) = 1;$$

$$B_5(\xi, \eta) = \frac{1}{6} - \frac{1}{2}\xi - \frac{\sqrt{3}}{6}\eta + \frac{1}{3}g_4 - \frac{2}{3}g_5 - \frac{1}{8}g_6 - \frac{1}{8}g_7 + \frac{1}{8}g_8 + \frac{1}{8}g_9,$$

$$\partial B_5/\partial \xi(P_2) = 1;$$

$$B_6(\xi, \eta) = -\frac{1}{6} - \frac{1}{2}\xi + \frac{\sqrt{3}}{6}\eta + \frac{2}{3}g_4 - \frac{1}{3}g_5 - \frac{1}{4}g_6 + \frac{1}{8}g_7 + \frac{1}{8}g_8 + \frac{1}{8}g_9,$$

$$\partial B_6/\partial \xi(P_3) = 1;$$

$$B_7(\xi, \eta) = -\frac{\sqrt{3}}{9} - \frac{2}{3}\eta + \frac{\sqrt{3}}{9}g_4 + \frac{\sqrt{3}}{9}g_5 + \frac{5\sqrt{3}}{24}g_6 - \frac{\sqrt{3}}{6}g_7 - \frac{5\sqrt{3}}{24}g_8$$
$$- \frac{\sqrt{3}}{9}g_9,$$

$$\partial B_7/\partial \eta(P_1) = 1;$$

$$B_8(\xi, \eta) = \frac{\sqrt{3}}{18} - \frac{\sqrt{3}}{6}\xi - \frac{1}{6}\eta + \frac{\sqrt{3}}{9}g_4 + \frac{\sqrt{3}}{9}g_5 - \frac{\sqrt{3}}{24}g_6 - \frac{\sqrt{3}}{24}g_7$$
$$+ \frac{\sqrt{3}}{12}g_8 - \frac{7\sqrt{3}}{24}g_9,$$

$$\partial B_8/\partial \eta(P_2) = 1;$$

$$B_9(\xi, \eta) = \frac{\sqrt{3}}{18} + \frac{\sqrt{3}}{6}\xi - \frac{1}{6}\eta + \frac{\sqrt{3}}{9}g_4 + \frac{\sqrt{3}}{9}g_5 + \frac{\sqrt{3}}{12}g_6 - \frac{7\sqrt{3}}{24}g_7$$
$$- \frac{\sqrt{3}}{24}g_8 - \frac{\sqrt{3}}{24}g_9,$$

$$\partial B_9/\partial \eta(P_3) = 1.$$

An interpolant on the standard triangle is now determined by the (transformed) nodal values and cardinal functions in the following form:

$$u(\xi, \eta) = \sum_{j=1}^{3} u(P_j)B_j(\xi, \eta) + \sum_{j=1}^{3} \left(\frac{\partial u}{\partial z_j}\right)(P_j)B_{3+j}(\xi, \eta)$$

$$+ \sum_{j=1}^{3} \left(\frac{\partial u}{\partial w_j}\right)(P_j)B_{6+j}(\xi, \eta) + \sum_{k=4}^{6} \left(\frac{\partial u}{\partial n_k}\right)(P_{3+k})B_{6+k}(\xi, \eta).$$

Let us make two comparisons. This triangular element can be compared with the bicubic rectangle. In both cases a regular mesh of vertices is required. For the triangle there is the disadvantage of handling the finer subdivision of triangles into the subtriangles formed by the seams. However, it has two advantages that may be very significant: It is piecewise quadratic (rather than bicubic), and it requires only function-value and first-derivative nodes.

In making comparisons with the Clough–Tocher triangle, let us suppose that the "averaging" idea is used to avoid use of the mid-side-derivative nodes. In this case the two can be used with precisely the same nodal configuration. Also, the two triangular elements will reproduce all quadratic functions but not all cubics. The present element is piecewise quadratic, whereas the Clough–Tocher element is piecewise cubic but with fewer subtriangles. The main advantage of the Clough–Tocher triangle is, of course, the fact that it can be applied on arbitrary triangulations.

There are other regular triangulations on which the piecewise quadratic strategy is successful. One is illustrated in Fig. 9.19 in which rectangular

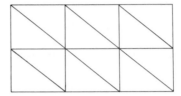

FIG. 9.19   Regular triangular mesh.

mesh is uniformly subdivided by diagonals (with negative slopes). However, one must use an appropriately chosen standard element (Fig. 9.20); one of equilateral shape will not do. The right-angled triangle with vertices at $(0, 0)$, $(1, 0)$, and $(0, 1)$ is appropriate. The seams of the triangle are segments of the lines $\eta + 2\xi = 1$, $\xi = \eta$, and $2\eta + \xi = 1$.

A cardinal basis is again constructed in two stages. First, we define

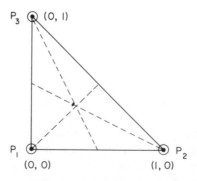

FIG. 9.20   Standard piecewise quadratic triangle.

$h_1, \ldots, h_9$ on the standard triangle by

$$h_1(\xi, \eta) = 1, \qquad\qquad h_2(\xi, \eta) = \xi,$$

$$h_3(\xi, \eta) = \eta, \qquad\qquad h_4(\xi, \eta) = (\eta - \xi)_+^2,$$

$$h_5(\xi, \eta) = (\xi - \eta)_+^2, \qquad h_6(\xi, \eta) = (1 - 2\xi - \eta)_+^2,$$

$$h_7(\xi, \eta) = (2\xi + \eta - 1)_+^2, \qquad h_8(\xi, \eta) = (1 - \xi - 2\eta)_+^2,$$

$$h_9(\xi, \eta) = (\xi + 2\eta - 1)_+^2.$$

Then, the nine cardinal functions are as follows:

$$B_1(\xi, \eta) = \frac{5}{3} - 2\xi - 2\eta - \frac{1}{3}h_4 - \frac{1}{3}h_5 - \frac{1}{3}h_6 + \frac{2}{3}h_7 - \frac{1}{3}h_8 + \frac{2}{3}h_9,$$

$$B_1(0, 0) = 1,$$

$$B_2(\xi, \eta) = -\frac{1}{3} + 2\xi + \frac{2}{3}h_4 - \frac{1}{3}h_5 + \frac{2}{3}h_6 - \frac{1}{3}h_7 - \frac{1}{3}h_8 - \frac{1}{3}h_9,$$

$$B_2(1, 0) = 1,$$

$$B_3(\xi, \eta) = -\frac{1}{3} + 2\eta - \frac{1}{3}h_4 + \frac{2}{3}h_5 - \frac{1}{3}h_6 - \frac{1}{3}h_7 + \frac{2}{3}h_8 - \frac{1}{3}h_9,$$

$$B_3(0, 1) = 1,$$

$$B_4(\xi, \eta) = \frac{5}{18} - \frac{1}{3}\xi - \frac{1}{3}\eta - \frac{1}{18}h_4 - \frac{1}{18}h_5 - \frac{7}{18}h_6 + \frac{1}{9}h_7 + \frac{1}{9}h_8 + \frac{1}{9}h_9,$$

$$\partial B_4 / \partial \xi(0, 0) = 1,$$

$$B_5(\xi, \eta) = \frac{1}{9} - \frac{2}{3}\xi - \frac{2}{9}h_4 + \frac{5}{18}h_5 - \frac{2}{9}h_6 + \frac{5}{18}h_7 + \frac{1}{9}h_8 + \frac{1}{9}h_9,$$

$$\partial B_5 / \partial \xi(1, 0) = 1,$$

$$B_6(\xi, \eta) = -\frac{1}{18} + \frac{1}{3}\eta - \frac{7}{18}h_4 + \frac{1}{9}h_5 - \frac{1}{18}h_6 - \frac{1}{18}h_7 + \frac{1}{9}h_8 + \frac{1}{9}h_9,$$

$$\partial B_6 / \partial \xi(0, 1) = 1$$

$$B_7(\xi, \eta) = \frac{5}{18} - \frac{1}{3}\xi - \frac{1}{3}\eta - \frac{1}{18}h_4 - \frac{1}{18}h_5 + \frac{1}{9}h_6 + \frac{1}{9}h_7 - \frac{7}{18}h_8 + \frac{1}{9}h_9,$$

$$\partial B_7 / \partial \eta(0, 0) = 1,$$

$$B_8(\xi, \eta) = -\frac{1}{18} + \frac{1}{3}\xi + \frac{1}{9}h_4 - \frac{7}{18}h_5 + \frac{1}{9}h_6 + \frac{1}{9}h_7 - \frac{1}{18}h_8 - \frac{1}{18}h_9,$$

$$\partial B_8 / \partial \eta(1, 0) = 1,$$

$$B_9(\xi, \eta) = \frac{1}{9} - \frac{2}{3}\eta + \frac{5}{18}h_4 - \frac{2}{9}h_5 + \frac{1}{9}h_6 + \frac{1}{9}h_7 - \frac{2}{9}h_8 + \frac{5}{18}h_9,$$

$$\partial B_9 / \partial \eta(0, 1) = 1.$$

The reader will easily determine the affine transformation of triangles of the mesh onto the standard triangle and hence the necessary scaling of derivative nodes.

We may ask why the negative sloping diagonals are used in Fig. 9.19 rather than those of positive slope. Indeed, the same data set can be used to generate two (possibly different) interpolants, depending on the choice of triangulation. This is illustrated in Figs. 9.21 and 9.22, in which the same

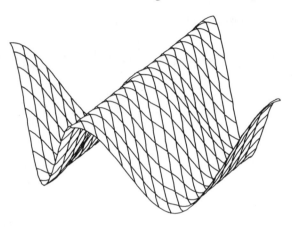

FIG. 9.21   Piecewise quadratic interpolant on 2×2 element rectangular grid, triangulated via positive sloping diagonals. Interpolated data at grid points: unit function values at five points, zero derivatives at all nine points.

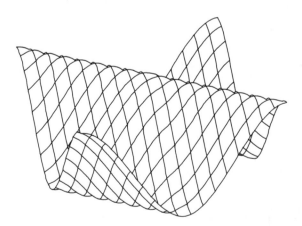

FIG. 9.22   Piecewise quadratic interpolant on 2×2 element rectangular grid, triangulated via negative sloping diagonals. Interpolated data as in Fig. 9.21.

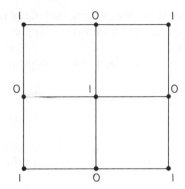

FIG. 9.23   Function value nodes for Figs. 9.21 and 9.22.

data set is used. On a regular lattice of nine points (Fig. 9.23), the function value nodes are as indicated, and *all* first derivative nodes are set equal to zero. We see that completely different surfaces are generated by the two choices of diagonal subdivision. This asymmetry is removed in our last example of a finite element.

9.9 A piecewise quadratic rectangular element

We take advantage of the discussion of the last paragraphs of Section 9.8 to determine an interpolant in the following way. The rectangle has the nodal configuration of Fig. 9.24, and we suppose the 12 nodal values are given.

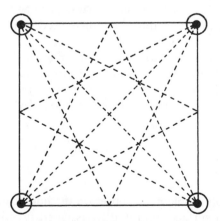

FIG. 9.24   Twelve-node piecewise quadratic element.

Step 1.   Using the negative sloping diagonal of the rectangle to determine two triangles, use the piecewise quadratic triangles of Section 9.8 to find an interpolant $u_1(\xi, \eta)$.

Step 2.   Repeat step 1 using the positive sloping diagonal to obtain an interpolant $u_2(\xi, \eta)$.

Step 3.   Since $\frac{1}{2}(u_1 + u_2)$ will again be an interpolant, we define

$$u(\xi, \eta) = \tfrac{1}{2}(u_1(\xi, \eta) + u_2(\xi, \eta)).$$

Note that the averaging of step 3 removes the bias associated with the choice of diagonal. To illustrate, Fig. 9.25 shows the averaged interpolant

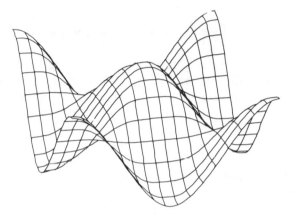

FIG. 9.25   Averaged piecewise quadratic interpolant on 2×2 element rectangular grid. Interpolated data as in Figure 9.23.

for the data of Fig. 9.23. This should be compared with Figs. 9.21 and 9.22. Another example in which this element is used can be found in Section 9.11.

In effect, the averaging procedure determines a single piecewise-quadratic rectangular finite element, although, as indicated in Fig. 9.24, the configuration of the seams is now complex. A related and more efficient piecewise quadratic element has been developed by Sibson and Thomson (1984).

9.10 Triangulation

Many interpolation problems, whether on the line or in the plane, require that a curve, or a surface, be fitted to assigned function values at nodes that are irregularly spaced. When the interpolation is on the line, this presents no great difficulties, but if the data points are scattered in the plane, the situation is not so clear. One useful way to handle this problem is to construct a triangulation of the region formed by the convex hull of the data points in such a way that the set of triangle vertices is precisely the set of data points. The purpose of this section is to indicate a "natural" way of doing this. Once it is done, a finite element method using general triangular elements may be used to generate an interpolating surface. The simplest of these is, of course, the surface composed of plane triangular faces, as described in Section 9.4. If derivative data is available at the vertices (or can be "boot-strapped" from the function value data), then the possibility arises of using the Clough–Tocher triangle of Section 9.6 to generate a $\mathscr{C}^1$ interpolating surface.

Returning to the triangulation problem it is first clear that, even with as few as four points, more than one triangulation may be possible, and a technique is needed to select a "good" triangulation. Intuitively, one anticipates that the goodness of a triangle is measured by its proximity to an equilateral triangle. Thus, the triangulation should eschew long, thin, and obtuse triangles as much as possible.

There is a technique with this property which has a long history and which has been rediscovered several times. First, it is convenient to approach a dual problem. We are given a set of N points, $P_1, \ldots, P_N$, scattered through the plane and we wish to divide the plane into N regions, $R_1, R_2, \ldots, R_N$. For $j = 1, 2, \ldots, N$, R_j consists of all the points of the plane that are closer to P_j than to any of the other $N - 1$ points. Hence, R_j contains P_j itself and no other given point. It turns out that each R_j is a (possibly infinite) convex polygon.

To construct R_j, we have only to form the perpendicular bisectors of the lines $P_j P_k$, where $k = 1, 2, \ldots, N$ and $k \neq j$. These lines will determine a unique polygon containing P_j in its interior with the defining property of R_j. The sides of each polygon are segments (or half-lines) of these perpendicular bisectors. Following this construction for each j, we generate the regions $R_1, R_2, \ldots, R_N$, and (the closures of) these regions are found to fill the whole plane.

These regions are variously associated with the names of Dirichlet, Voronoi, and Thiessen. We shall call them a *Dirichlet tessellation* of the plane. Their position in a more general mathematical context is described by Rogers (1964). Rhynsburger (1973) gives a useful introduction to the theory and application. Green and Sibson (1978) seem to develop the first algorithmic approach to finding the Dirichlet tessellation, and a more recent reference is the paper by Cline and Renka (1984).

Once the Dirichlet tessellation is found, a corresponding triangulation of the convex hull of $P_1, P_2, \ldots, P_N$ is obtained by joining the point P_j to each other point P_k for which R_j and R_k share an edge. This procedure is found to generate a triangulation of the convex hull of $P_1, P_2, \ldots, P_N$. In Fig. 9.26 we illustrate the Dirichlet tessellation and corresponding triangulation for a set of nine points.

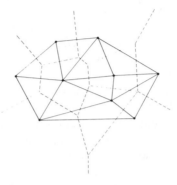

FIG. 9.26   Dirichlet tessellation and corresponding triangulation.

To support our claim that this construction generates triangulations that tend to favour acute triangles, consider the two triangulations of the same four points in Fig. 9.27. The reader should carry through the construction via a Dirichlet tessellation and confirm that the configuration of Fig. 9.27a results.

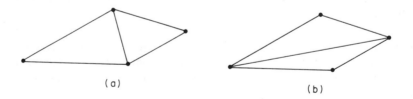

(a)   (b)

FIG. 9.27   Two triangulations of four points.

9.11 Approximation of a topographic map

In this section we present a more ambitious example and use it to illustrate three of the techniques developed in Chapters 8 and 9. A topographic map on the scale of $1:50,000$ was selected which represents a 12 km × 8 km rectangle of a mountainous region of central British Columbia (Fig. 9.28). Sections of the region were constructed along longitudinal and latitudinal grid lines at 1 km spacing. Thus, 21 such sections were developed as

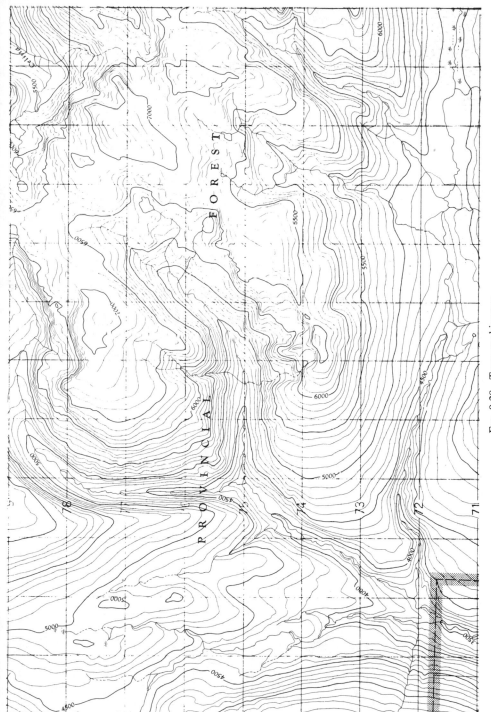

FIG. 9.28   Topographic map.

FIG. 9.29 Spline blended representation.

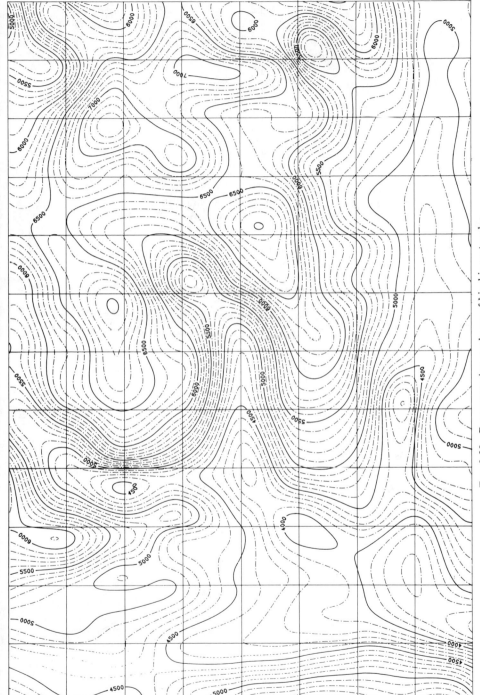

FIG. 9.30 Representation made up of bicubic rectangles.

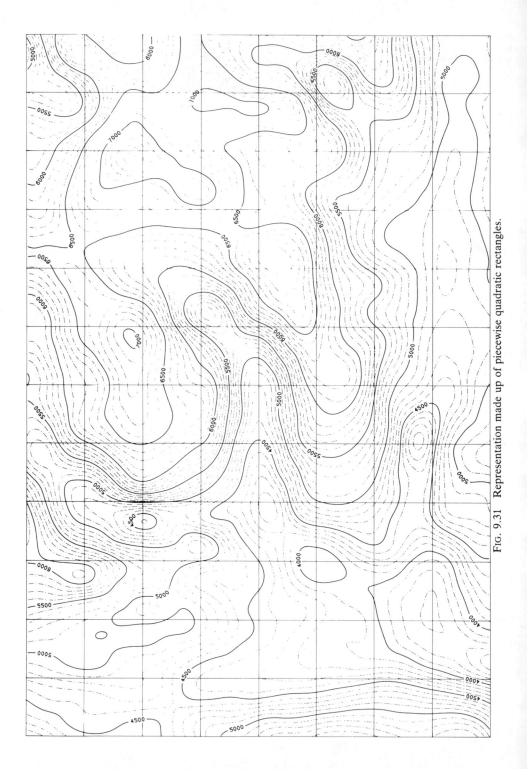

FIG. 9.31   Representation made up of piecewise quadratic rectangles.

accurately as possible by a curve-fitting method using $\mathscr{C}^1$ cubic splines which minimize a weighted norm of the second derivative of the spline (cf. Section 4.4). These sections were then interpolated to generate smooth surfaces, as described next in Examples 1–3. Constructing a surface from sections, as we do in these examples, could be seen as modelling procedures commonly used in seismic surveys in which data is generated on sections by acoustic or electromagnetic techniques.

EXAMPLE 1   Our first procedure is to use spline blending as described in Section 8.5 and especially Theorem 8.5.1. A contour map of the result is presented in Fig. 9.29. Recall that, in effect, heights on all the sections reproduce the heights of the topographic map accurately (although some smoothing inevitably occurs in generating our sections). So the errors produced (which are apparent) result mainly from the surface-fitting procedure used to "fill-in" a smooth surface between the grid lines. Although some of the main surface features are reproduced well, it is clear that the surface-fitting procedure has real difficulty in reproducing the dendritic character of the region.

EXAMPLE 2   In the second procedure we utilize the bicubic rectangle interpolant as described in Sections 7.6, 8.4 (Example 4), and 9.3. In this case, each 1 km × 1 km square is treated as a finite element, and the 16 nodal values required for the interpolating surface are obtained from the sections of the region described above. The nodal values for heights and slopes at the vertices were obtained directly from those sections and the cross-derivatives estimated by an averaging technique. The resulting $\mathscr{C}^1$ surface is illustrated in Fig. 9.30. As one might expect, the fact that we do not interpolate along entire sections, but only use data at the vertices, means that we lose a lot of information about the original, and the result is a poorer representation than that of Example 1.

EXAMPLE 3   This example is produced mainly for comparison with Example 2. It is a similar technique but, instead of using the bicubic rectangle with each of the 1 km × 1 km squares, the piecewise quadratic element of Section 9.9 is employed. In this case the cross-derivative nodes of the bicubic rectangle are not required. A contour diagram of the resulting surface appears as Fig. 9.31. As one might expect, the repesentations of Examples 2 and 3 are very much alike.

10

Moving Least Squares and Composite Methods

10.1 Introduction

The basic problem which motivates this chapter is that of surface fitting to function values assigned at data points that are scattered in the plane. A moving least squares procedure is discussed first. This is a natural generalization of the corresponding techniques for curve fitting discussed in some detail in Sections 2.9–2.11, especially when combined with the orthodox least squares surface-fitting method of Section 7.7. Indeed, our exposition will rely heavily on the earlier discussions.

As in one space dimension, it will be found that *interpolants* can be generated by moving least square techniques, so we shall use the IMLS methods once more. These will be seen to provide a class of complete solutions to the problem of interpolating smooth functions to scattered data. However, it will also be seen that these solutions are computationally expensive and (a lesser criticism) subject to unfortunate edge effects. Consequently, IMLS methods are more frequently used in surface-fitting problems in combination with other methods. For example, if the scattered data points are used as the vertices of a triangulation as outlined in Section 9.10, then it will be possible to find the gradient of the IMLS surface at each data point. There would then be sufficient information to implement a finite element procedure (e.g. the Clough–Tocher triangle), which will determine a $\mathscr{C}^1$ interpolating surface. Alternatively, the IMLS surface may be sampled at a *regular* lattice of points on which a surface can be constructed by tensor product or by other finite element methods. These are among the "composite" methods to be discussed later in Sections 10.5 and 10.6.

10.2 Weighted least squares fitting

It will be convenient to abbreviate the notation (x, y) for a point in the plane to z. Thus, we shall suppose that the (generally scattered) distinct data points in the plane are given by $z_i = (x_i, y_i)$, for $i = 1, 2, \ldots, N$. If the function value f_i is given at z_i for each i, a function u is to be found, which is defined on a domain containing the convex hull of $z_1, z_2, \ldots, z_N$ with the properties that u is $\mathscr{C}^1$ on this domain and $u(z_i) = f_i$ for $i = 1, 2, \ldots, N$.

For this discussion it is easier (from the point of view of notation) to consider more general least squares techniques than those based on polynomials. We shall suppose that n linearly independent functions are given and are defined on the whole plane. We call them $b_1(z), b_2(z), \ldots, b_n(z)$ and assume that $n \leqslant N$; usually, n is very much less than N. For least squares approximation with bivariate quadratic polynomials, we would take $n = 6$, and the functions would be simply,

$$b_1(z) = 1, \quad b_2(z) = x, \quad b_3(z) = y,$$
$$b_4(z) = x^2 \quad b_5(z) = xy, \quad b_6(z) = y^2, \tag{10.1}$$

and similarly for fitting with polynomials of other degrees. There are situations in which exponential or trigonometric functions, for example, can be used for $b_1, b_2, \ldots, b_n$, but in the great majority of applications these are likely to be the monomials, which form a basis for a space of bivariate polynomials (see Section 7.2).

We now assume that the fitted surface will be a linear combination of the chosen functions $b_1, b_2, \ldots, b_n$. Thus, we assume initially that, for any point z,

$$u(z) = \sum_{j=1}^{n} a_j b_j(z), \tag{10.2}$$

for some choice of the numbers $a_1, a_2, \ldots, a_n$, which is independent of z and is to be determined.

As in the formulation of Eq. (7.12), the fit of such a function $u(z)$ to the data values $f_1, \ldots, f_N$ can be measured by the size of the error functional

$$E(u) = \sum_{i=1}^{N} (u(z_i) - f_i)^2 = \sum_{i=1}^{N} \left(\sum_{j=1}^{n} a_j b_j(z_i) - f_i \right)^2. \tag{10.3}$$

We take a more general position and suppose that the separate terms in the summation over i are given different weights $w_i > 0$. Thus, we consider the functional

$$E(u) = \sum_{i=1}^{N} w_i (u(z_i) - f_i)^2. \tag{10.4}$$

Of course, we recover Eq. (10.3) by setting $w_1 = w_2 = \ldots = w_N = 1$, and it is found that the final fitted surface is drawn closer to f_i at those points z_i with a relatively large weight w_i.

As in Section 7.7, the function u that minimizes E is obtained by first finding the coefficients $a_1, a_2, \ldots, a_n$ of Eq. (10.2), and these coefficients are found by solving the normal equations $\partial E / \partial a_i = 0$ for $i = 1, 2, \ldots, n$. It is not difficult to verify that these equations can be written in the following matrix–vector form. Define the $n \times N$ matrix B whose jth row is $[b_j(z_1), b_j(z_2), \ldots, b_j(z_N)]$, for $j = 1, 2, \ldots, n$, and an $N \times N$ diagonal matrix $W = \mathrm{diag}[w_1, w_2, \ldots, w_N]$. Define column vectors $\mathbf{a}$ and $\mathbf{f}$, the first with elements $a_1, a_2, \ldots, a_n$ (to be found) and the second with the data $f_1, f_2, \ldots, f_n$ as elements. Note that the matrix product BWB^T is then an $n \times n$ matrix. Furthermore, the normal equations have the form,

$$BWB^T\mathbf{a} = BW\mathbf{f}. \tag{10.5}$$

These equations then have a unique solution and can be solved by standard methods as long as the matrix BWB^T is nonsingular (see the remarks and references at the end of Section 2.6). Note that with the choice [Eq. (10.1)] of basis functions, the matrix B can be identified with matrix V^T of Eq. (7.14). Also, in Eq. (7.14) we have $W = I$.

We conclude that on solving the n simultaneous linear equations represented by Eq. (10.5), the coefficients $a_1, a_2, \ldots, a_n$ are obtained, and the function $u(z)$ that minimizes the error functional $E(u)$ of Eq. (10.4) is obtained on substitution into Eq. (10.2).

This technique may sometimes be useful in its own right, but our objective is to use this discussion in the development of *moving* least squares methods. This is done next in Section 10.3.

10.3 Moving least squares methods

Let us return in the discussion of Section 10.2 to the formulation of the error functional $E(u)$ of Eq. (10.4). As in Section 2.9 it can be argued that the value of u at a point z should be most strongly influenced by the values of f_i at those points z_i that are closest to z. This implies that weights $w_1, w_2, \ldots, w_N$, which should be assigned in Eq. (10.4), are dependent on z and have the property that $w_i(z)$ decreases in magnitude as the distance from z to z_i increases. Then, Eq. (10.4) is replaced by

$$E_z(u) = \sum_{i=1}^{N} w_i(z)(u(z_i) - f_i)^2. \tag{10.6}$$

Holding z fixed, the analysis proceeds as before to obtain normal equations of the form

$$BW(z)B^T\mathbf{a}(z) = BW(z)\mathbf{f}. \tag{10.7}$$

Here, B and $\mathbf{f}$ are as before; the essential difference is that

$$W(z) = \text{diag}[w_1(z), w_2(z), \ldots, w_N(z)],$$

is now z-dependent so that the solution $\mathbf{a}$ is also z-dependent. From Eq. (10.2) we have now

$$u(z) = \sum_{j=1}^{n} a_j(z)b_j(z). \tag{10.8}$$

The great disadvantage of this process is, of course, that a set of normal equations [Eq. (10.6)] must be solved for every value of z at which the surface height is to be calculated and, in addition, the weights $w_1, \ldots, w_N$ must be recalculated in a systematic way for each z.

The Euclidean distance between points $z = (x, y)$ and $\hat{z} = (\hat{x}, \hat{y})$ is given by

$$d(z, \hat{z}) = \sqrt{(x - \hat{x})^2 + (y - \hat{y})^2}. \tag{10.9}$$

Since the weighting scheme is to depend only on the distance between points in the plane, the weights $w_1(z), \ldots, w_N(z)$ will depend on one function $w(d)$. Thus, they are determined by a non-increasing function $w(d)$ by writing

$$w_i(z) = w(d(z, z_i)). \tag{10.10}$$

As in Section 2.9, there are many candidates for the function w, and the criteria for determining a suitable w are much the same. The important questions are the behaviour as w as $d \to \infty$ and as $d \to 0$.

First, if the procedure is to *interpolate* at all the data points, then we must have $w \to \infty$ as $d \to 0$. This suggests that, at least for small values of d, $w(d)$ should behave like d^{-2} or d^{-4}. For large values of d, w must attenuate rapidly enough to minimize, or remove entirely, the influence of remote data values. The arguments for truncation when N is very big (as in Section 2.9) are persuasive. The two-dimensional analogues of Eqs. (2.29) and (2.30) are easily formulated.

McLain (1974) recommends the use of a function such as

$$w(d) = e^{-d^2}/d^2, \tag{10.11}$$

which clearly has the correct asymptotic behaviour at $d = 0$ and $d = \infty$. In numerical practice the singularity at $d = 0$ may present some difficulties when d is small. In this case it has been proposed that the function $w(d) = d^{-2}$, for example, may be replaced by $w(d) = (d^2 + \varepsilon)^{-1}$, for some small fixed $\varepsilon > 0$. Indeed, ε should be so small that, to numerical and visual accuracy, interpolation is achieved at $d = 0$.

The weight function in Eq. (10.11) can be similarly modified in an effort to control overflow. As well, the exponent can be changed to alter the rate of attenuation in the exponential function. Indeed, the function

$$w(d) = e^{-\alpha d^2}/(d^2 + \varepsilon) \qquad (10.12)$$

with $\varepsilon = 0.001$ and $\alpha = 1/16$ was used in Fig. 10.1, 10.2, and with some examples of Section 10.5.

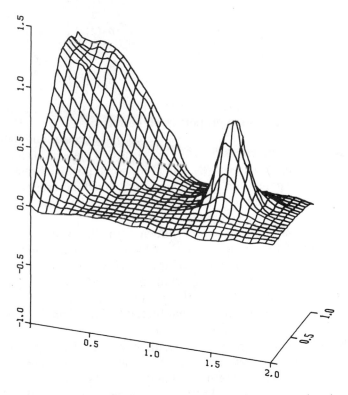

FIG. 10.1   Distance weighted least squares interpolant, perspective view.

If the objective is only to produce information on the slopes at the data points, it *is* possible to avoid the singular behaviour at $d = 0$. For example [following Lodwick and Whittle (1970)], another useful convention is simply to assume $w_i(z) = 1$ for the r data points closest to z (where $r > n$) and $w_i(z) = 0$ otherwise. Then the surface will not generally interpolate but may nevertheless give useful gradient approximations. Note that if

$r < n$, the coefficient matrix $BW(z)B^T$ of Eq. (10.7) will be singular, and the process will break down. If $r \geq n$ and is too "close" to n, there will be some risk that this matrix will be singular, or nearly so, for some values of z. Lodwick and Whittle (1970) use $r = 15$ when $n = 3$.

In general, if all the weights are positive, then the coefficient matrix $BW(z)B^T$ will be nonsingular for all z, provided there is one set of n data points on which the functions $b_1, b_2, \ldots, b_n$ are linearly independent. In other words, if $z_{i_1}, z_{i_2}, \ldots, z_{i_n}$ are the data points in question, the determinant

$$\det \begin{bmatrix} b_1(z_{i_1}) & b_1(z_{i_2}) & \ldots & b_1(z_{i_n}) \\ b_2(z_{i_1}) & b_2(z_{i_2}) & \ldots & b_2(z_{i_n}) \\ & & \vdots & \\ b_n(z_{i_1}) & b_n(z_{i_2}) & \cdots & b_n(z_{i_n}) \end{bmatrix},$$

is non-zero. We assume throughout that this is the case.

The smoothness and other properties of the surfaces generated by moving least squares methods have been investigated by Lancaster and Šalkauskas (1981). If the basis functions $b_1, b_2, \ldots, b_n$ are of class $\mathscr{C}^m$ (and they are generally $\mathscr{C}^\infty$, in fact), and if $w(d) = d^{-2k}$ for some positive integer k, then it is shown in that paper that the function $u(z)$, which determines the fitted surface, is also of class $\mathscr{C}^m$. In particular then, when $b_1, \ldots, b_n$ are polynomials, the surface generated is $\mathscr{C}^\infty$.

These techniques are also reproducing in the sense that, if the data corresponds to a function $b(z)$, which is a linear combination of $b_1(z), \ldots, b_n(z)$, then the result of the moving least squares process is simply $b(z) = u(z)$. This is an important property for many applications. For example, one may like to insist that the chosen procedure will reproduce all plane surfaces, and this can be guaranteed by including the functions 1, x, and y among $b_1, b_2, \ldots, b_n$.

The interest of the smoothness result concerns the behaviour of an IMLS surface near the data points because, at the points themselves, the coefficients of Eq. (10.7) are not defined. There is a singularity there in one of the elements of $W(z)$. The first step toward resolving this difficulty is to divide both sides of Eq. (10.7) by $\Sigma_{i=1}^N w_i(z)$. Since this is always positive, this operation does not affect the solution. Defining

$$v_j(z) = w_j(z) \Big/ \sum_{i=1}^N w_i(z), \qquad j = 1, 2, \ldots, N, \qquad (10.13)$$

and $V(z) = \text{diag}[v_1(z), \ldots, v_N(z)]$, Eq. (10.7) becomes

$$BV(z)B^T \mathbf{a}(z) = BV(z)\mathbf{f}. \qquad (10.14)$$

Now the functions $v_1(z), \ldots, v_N(z)$ are relatively well-behaved near the data points and can be defined by continuity at the data points themselves. We think of them as normalized weight functions and note the following properties:

(1) $v_i(z_j) = \delta_{ij}$ (the Kronecker delta) for $i, j = 1, 2, \ldots, N$;
(2) $0 \leqslant v_i(z) \leqslant 1$ for all z and $v_i(z) = 0$ if and only if $z = z_j$ and $j \neq i$;
(3) $\sum_{i=1}^{N} v_i(z) = 1$ for all z, and
(4) $v_i(z) \to N^{-1}$ as $d(z, 0) \to \infty$.

It follows immediately from properties 1 and 2 that, if $v_i(z)$ is differentiable at z_i, then the gradient of $v_i(z)$ at $z = z_j$ is zero for *any* j. That is, *all the directional derivatives of $v_i(z)$ evaluated at z_j are zero*. Consequently, $v_i(z)$ has a local maximum at z_i and a local minimum at each z_j, with $j \neq i$. The differentiability of $v_i(z)$ certainly follows if all the weight functions $w_j(z)$ with $j \neq i$ are differentiable at z_i. This is invariably the case.

This discussion applies immediately to the two-dimensional Shepard's method (see Section 2.11). In this case we take $n = 1$ and $b_1(z) \equiv 1$. Then the matrix B of Eq. (10.14) is simply,

$$B = [b_1(z_1), \quad b_1(z_2), \quad \ldots, \quad b_1(z_N)] = [1, \quad 1, \quad \ldots, \quad 1, \quad 1],$$

and so Eq. (10.14) reduces to

$$\left(\sum_{i=1}^{N} v_i(z) \right) a_1(z) = \sum_{i=1}^{N} v_i(z) f_i,$$

and the interpolant is given by Eq. (10.8). Noting property (3) of the normalized weight functions, we have simply,

$$u(z) = \sum_{i=1}^{N} f_i v_i(z). \tag{10.15}$$

In particular, we see that $v_i(z)$ is the ith cardinal function for Shepard's method and, since every such function displays the "flat-spot" phenomenon at all data points (i.e. all gradients there are zero), it follows from Eq. (10.15) that all Shepard interpolants have the same characteristic. This flat-spot phenomenon is one that may be useful in certain circumstances (see Gordon and Wixom (1978) for examples) but rules out Shepard's method as a general-purpose surface (or curve) interpolation scheme.

Concerning the choice of n, in extensive tests carried out by McLain (1974) polynomials with one, three, four, six, eight, ten, and fifteen terms are used. His recommendation, based on a trade-off of accuracy against computing time, is the use of the complete quadratic [Eq. (10.1)] together with a weighting function like Eq. (10.11) or Eq. (10.12).

To illustrate the general technique we return to our model problem (see Fig. 7.13). In Fig. 10.2 we indicate 150 randomly selected points in the domain of the model problem; thus, $N = 150$. The values of $f_1, f_2, \ldots, f_{150}$ are calculated from the analytically defined heights of the model surface [see Eq. (7.15)].

An IMLS interpolant was then computed using the six basis functions of Eq. (10.1) and the weight function of Eq. (10.12). Figure 10.2 shows a contour map while Fig. 10.1 is a perspective view of the interpolant.

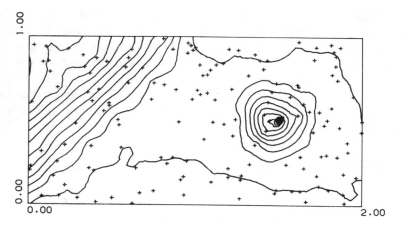

FIG. 10.2   Set of 150 random data points for model problem and contour map.

Note that this is, in fact, a $\mathscr{C}^\infty$ surface. There is an obvious coarseness superimposed by the plotting routine and the size of the mesh used for plotting. Anomalies at other points, especially on the edges, are largely due to the distribution and low density of data points in these regions.

10.4 Calculating derivatives on MLS surfaces

It has been remarked in Section 10.1 that, rather than using the IMLS surface itself, it is often useful simply to sample the height and the derivatives of such a surface at a coarse net of points and then to generate a surface by tensor product or finite element procedures. Before illustrating such composite methods, let us first consider the problem of computing the derivatives of a function $u(z)$ generated by an IMLS method, as described in Section 10.3. We shall see that the derivatives of $u(z)$ *at the data points* are relatively easy to find.

Recall that u is defined at any z by Eq. (10.8), where $a_1(z), \ldots, a_n(z)$ are the elements of the solution vector $\mathbf{a}(z)$ of Eq. (10.14). Introduce the column vector $\mathbf{b}(z)$ with elements $b_1(z), \ldots, b_n(z)$. Then Eq. (10.8) can be written in the form

$$u(z) = \mathbf{b}(z)^T\mathbf{a}(z). \tag{10.16}$$

To find the partial derivative of u in the direction of s, say, we differentiate as a product to obtain (using subscripts for partial derivatives):

$$u_s = \mathbf{b}_s^T\mathbf{a} + \mathbf{b}^T\mathbf{a}_s, \tag{10.17}$$

and all these functions are, of course, evaluated at the same point z.

Differentiating once more in the direction of a variable t, we obtain the second derivative

$$u_{st} = \mathbf{b}_{st}^T\mathbf{a} + \mathbf{b}_s^T\mathbf{a}_t + \mathbf{b}_t^T\mathbf{a}_s + \mathbf{b}^T\mathbf{a}_{st}. \tag{10.18}$$

In general, $\mathbf{b}(z)$ is a simple function and the calculation of its derivatives will pose no problem. The derivatives of $\mathbf{a}(z)$ are more troublesome and must be obtained from the defining Eq. (10.14). Differentiating both sides of Eq. (10.14) with respect to s yields

$$BV_sB^T\mathbf{a} + BVB^T\mathbf{a}_s = BV_s\mathbf{f},$$

so that $\mathbf{a}_s$ is obtained by solving

$$(BVB^T)\mathbf{a}_s = BV_s(\mathbf{f} - B^T\mathbf{a}), \tag{10.19}$$

keeping in mind that $\mathbf{a}$ itself is already determined by Eq. (10.14). One can, of course, take computational advantage of the fact that the coefficient matrices of Eqs. (10.14) and (10.19) are the same.

Consider, however, the right-hand side of Eq. (10.19) when evaluated at a data point. The matrix V_s is diagonal, and we have seen in Section 10.3 that for an IMLS method, all first-order derivatives of the functions $v_i(z)$, $i = 1, 2, \ldots, N$, vanish at any data points. Consequently, $V_s = 0$ at data points, and Eq. (10.19) implies $\mathbf{a}_s = \mathbf{0}$ also. Thus, in a more explicit notation, Eq. (10.17) reduces at the data points to

$$\partial u(z_j)/\partial s = [\partial\mathbf{b}^T(z_j)/\partial s]\mathbf{a}(z_j).$$

At all other points it is, of course, necessary to calculate $\partial\mathbf{a}/\partial s$ from Eq. (10.19) or the equivalent equation following from Eq. (10.7).

For second-order derivatives, it is found that

$$(BVB^T)\mathbf{a}_{st} = BV_{st}(\mathbf{f} - B^T\mathbf{a}) - BV_tB^T\mathbf{a}_s - BV_sV\mathbf{a}_t. \tag{10.20}$$

Combining the solution with first derivatives from Eq. (10.19), $\partial^2 u/\partial s\,\partial t$ is obtained from Eq. (10.18).

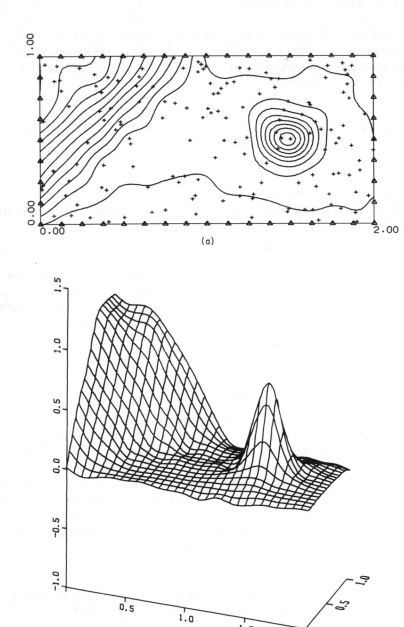

Fig. 10.3 Bicubic spline interpolant of IMLS surface. (a) Contour map and (b) perspective view.

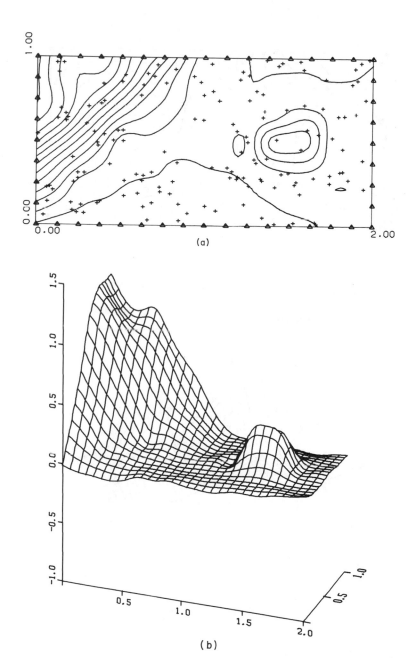

FIG. 10.4 Bicubic spline interpolant of IMLS surface. (a) Contour map and (b) perspective view.

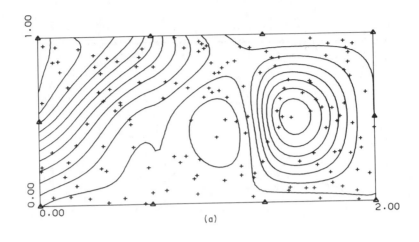

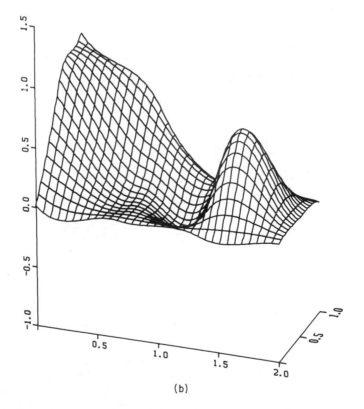

FIG. 10.5 Bicubic rectangle interpolant of IMLS surface; 3×2 grid. (a) Contour map and (b) perspective view.

When finding second-order derivatives at a data point, we will again have $\mathbf{a}_s = \mathbf{a}_t = \mathbf{0}$ so that Eqs. (10.18) and (10.20) reduce to

$$u_{st} = \mathbf{b}_{st}^T\mathbf{a} + \mathbf{b}^T\mathbf{a}_{st}, \qquad (BVB^T)\mathbf{a}_{st} = BV_{st}(\mathbf{f} - B^T\mathbf{a}).$$

10.5 Composition of the IMLS and bicubic spline methods

As our first illustrations of composite methods we consider the model problem of Fig. 7.13 once more and two sets of 150 randomly distributed data points. These appear as crosses in the lower parts of Figs. 10.3 and 10.4. (The points of Fig. 10.3 are the same as those of Fig. 10.1.)

Using the IMLS method with complete quadratic basis [as in Eq. (10.1)] and the weight function of Eq. (10.12), the IMLS surface height is sampled, in each case, on a uniform grid of $17 \times 9 = 153$ points. These points lie on the grid induced by the triangles shown on the edges of Figs. 10.3 and 10.4. A bicubic spline surface is then constructed which interpolates this data (see Sections 8.2 and 8.3). The perspectives and contour maps of these final composite surfaces are shown in Figs. 10.3 and 10.4. These should be compared with Fig. 8.5c in which the same data points are used in a bicubic spline interpolating surface but without the intermediate step of constructing the IMLS approximation.

Comparison of Fig. 10.3 with Fig. 10.1 is also interesting. This shows the smoothing effect of the bicubic surface on the IMLS surface, which it interpolates.

In comparison with our subsequent examples, the bicubic spline surface has the advantage that the calculation of derivatives of the IMLS surface (as described in Section 10.4) is not required. However, like subsequent examples based on a *regular* lattice of points, it has the disadvantage that, in general, it will no longer interpolate to the data at the original 150 random points.

10.6 Composition of the IMLS and bicubic rectangle methods

In this section the IMLS method is first implemented as in Section 10.5 and based on the 150 random data points of Fig. 10.2. Next, the bicubic rectangle is used (as a finite element technique) to approximate the IMLS surface. First, we subdivide the domain of the test problem into $3 \times 2 = 6$ rectangles. Thus, we sample only 12 distinct points of the IMLS surface. However, we need four pieces of data at each point f, f_x, f_y, and f_{xy}. The derivative data are computed as described in Section 10.4. The resulting $\mathscr{C}^1$ surface is illustrated in Fig. 10.5.

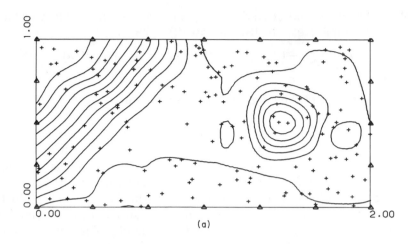

(a)

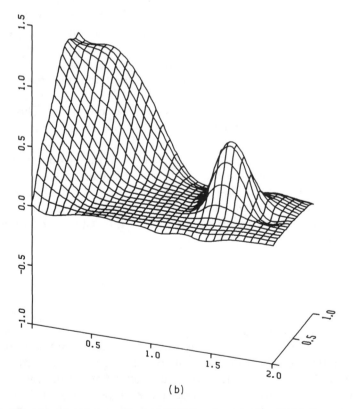

(b)

FIG. 10.6 Bicubic rectangle interpolant of OMLS surface; 6 × 4 grid. (a) Contour map and (b) perspective view.

The procedure adopted for the surface of Fig. 10.6 is the same but on a finer grid. There are $6 \times 4 = 24$ rectangles and hence 35 points and 140 nodal values.

These surfaces should be compared with those of Figs. 9.7 and 9.8 where the data are read directly from the model surface, without reference to the IMLS surface. Note that the bicubic rectangle requires the calculation of second-order derivatives and hence the implementation of an equation like Eq. (10.20) to find these derivatives.

10.7 Composition of the IMLS and seamed-triangle methods

Calculation of the second-order derivatives required by the bicubic rectangle can be avoided with use of the Clough–Tocher triangle. Here, we do not use the full power of the Clough–Tocher triangle. Although it *can* be used on a general triangulation, it is used here on a set of regular isosceles triangles, as illustrated in Fig. 10.7. To avoid difficulties with edge effects, interpolation is completed on the 18 isosceles triangles shown in Fig. 10.7, and the results are presented only on the truncated domain indicated by the dotted lines of Fig. 10.7. To reduce computation time, the

FIG. 10.7   Isosceles triangulation.

mid-side nodes are obtained by averaging the corresponding directional derivatives at the two associated vertices (see Section 9.7). Thus, it is necessary to compute from the IMLS surface just three nodal values at each vertex, making a total of 48 nodes. The resulting composite surface is shown in Fig. 10.8.

Comparison with Fig. 10.5 is interesting. The same number of nodal values is used in each case to produce a $\mathscr{C}^1$ surface. We obtain a similar resolution in the two cases, but the present method is easier to implement since no second derivative nodes are required. However, this method has missed most of the mountain.

A refined tessellation of 65 complete isosceles triangles (5 rows of 13 triangles) produces 45 vertices and 135 nodes. This is used in a similar way to obtain Fig. 10.9. This result compares favourably with Fig. 10.6 which required a similar number of nodal values and use of the bicubic rectangle.

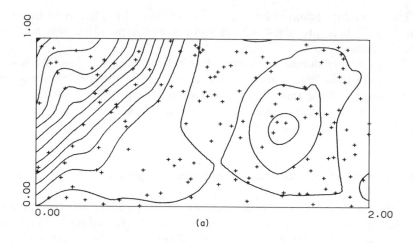

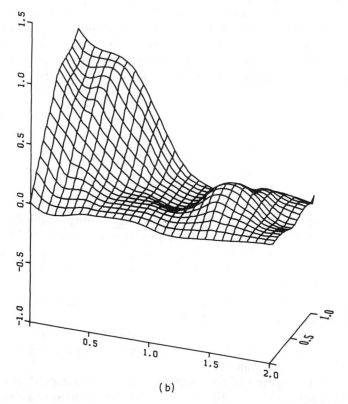

FIG. 10.8   Clough–Tocher interpolant of IMLS surface; 48 nodes. (a) Contour map and (b) perspective view.

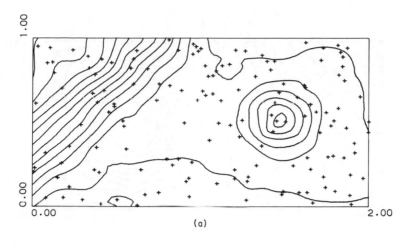

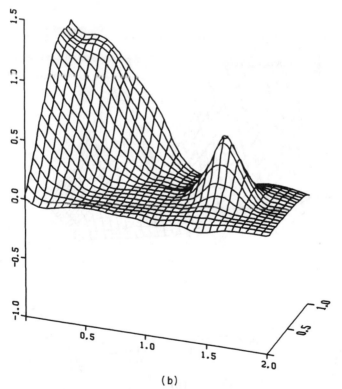

FIG. 10.9   Clough–Tocher interpolant of IMLS surface, 135 nodes. (a) Contour map and (b) perspective view.

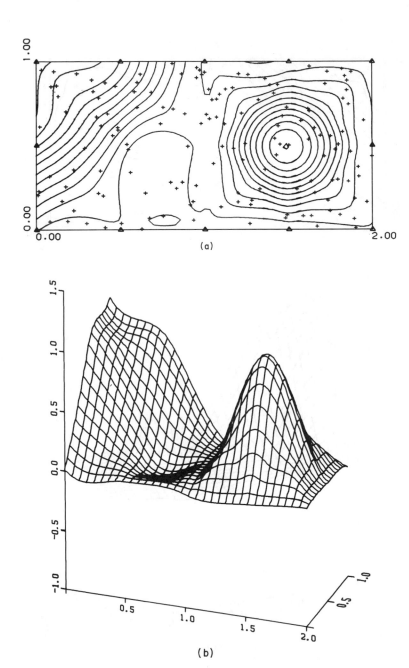

FIG. 10.10   Seamed rectangle interpolant of IMLS; 45 nodes. (a) Contour map and (b) perspective view.

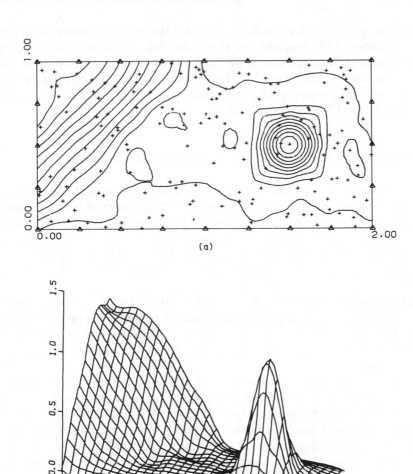

FIG. 10.11   Seamed rectangle interpolant of IMLS; 135 nodes. (a) Contour map and (b) perspective view.

Now recall the seamed quadratic triangle interpolation scheme of Section 9.8. Because of the regularity of the triangulations considered here, this can be used in the same way as the Clough–Tocher triangle and using just the same nodal values. The results are, in fact, visually much alike but the surface is now made up of (more) quadratic patches rather than the cubic patches of the Clough–Tocher triangle.

10.8 Composition of the IMLS and seamed-rectangle methods

Now consider a rectangular finite element scheme based on the "averaged" piecewise quadratic element of Section 9.9. The number of rectangles used in this procedure is adjusted so that the total numbers of nodes admit comparison with the surfaces generated using the bicubic rectangle, as described in Section 10.6. Since the quadratic rectangle requires three nodes per vertex, a rectangular grid of $4 \times 2 = 8$ rectangles produces a total of 15 vertices and 45 nodal values.

The resulting surface is illustrated in Fig. 10.10. The extra labour involving in "averaging" two separate solutions may be justifiable on the grounds that the cardinal functions involved are less complicated than those in the other schemes used here, and the resolution of the model surface seems to be improved.

A refined rectangular mesh of $8 \times 4 = 32$ rectangles produces 45 vertices and 135 nodal values. The corresponding surface is presented in Fig. 10.11. Again, this compares favourably with Figs. 10.4, 10.6 and 10.8, each of which is obtained using a similar number of nodal values and the same underlying IMLS approximation to the model surface. However, the use of a rectangular element results in some boxiness in the contour lines defining the mountain. This effect is somewhat less pronounced in Figs. 10.5 and 10.6.

11

Surface Splines

11.1 Introduction

Before embarking on the topic of this chapter, it is helpful to review some aspects of the linear and cubic splines discussed in Chapters 3 and 4.

In the case of linear splines, one interpretation of them is as linear segments, each defined on a subinterval $[k_i, k_{i+1}]$ created by a partition of the domain by knots $a = k_0 < k_1 < \ldots < k_N = b$ and joined together by the requirement of continuity at the knots. In contrast to this piecewise interpretation, we also considered linear splines as linear combinations of the functions $|x - k_i|$, which are smooth everywhere but at $x = k_i$, where they have a corner and thus fail to possess a first derivative. One may like to think of a straight-line function, which is too rigid for use in interpolation, being loosened by permitting non-differentiability at just one point. Seen from this point of view, it is not the piecewise nature of linear splines which is paramount; rather, the location of non-differentiabilities (or singularities) is the dominant feature.

Cubic splines can also be viewed from two directions. As piecewise cubics, they are third-degree polynomial segments connected by the requirement of continuity of the spline and its first and second derivatives at the knots. On the other hand, they can be constructed from linear combinations of the $\mathscr{C}^2$ functions $|x - k_i|^3$ which fail to be $\mathscr{C}^\infty$ only at the knots, together with the functions 1 and x. This is the approach taken in Section 3.6.

When we come to surface interpolation at scattered points in the plane, the two points of view can be generalized. The piecewise approach leads to the consideration of finite elements (Chapter 9), the domain being subdivided by lines joining the knots in some rectangular or triangular pattern. The second line of attack leads us to employ translates of a function usually with some non-differentiability at the knots as a basis for interpolation. This method follows almost verbatim the approach taken in

Chapter 5 (Section 5.4 et seq.), where Boolean sums are employed in order to ensure that even when the basis of translates does not contain polynomials, low-degree polynomial data will be correctly interpolated.

11.2 Interpolation with translates

In Chapter 5 we indicated the value of choosing as an interpolant the linear combination $\sum_{i=1}^{N} a_i \varphi(x - x_i)$ of translates $\varphi(x - x_i)$ of a *basic function* $\varphi(x)$. This results in an interpolant whose shape is independent of the position of the origin on the (horizontal) x-axis. We should do the same in the case of surface interpolation. As well, the *direction* of the horizontal axis will not affect the form of the interpolant, provided that $\varphi(-x) = \varphi(x)$. In the bivariate case this means that φ should have *rotational symmetry*. That is, that φ should take the same value at all points of a circle with centre at the origin of coordinates.

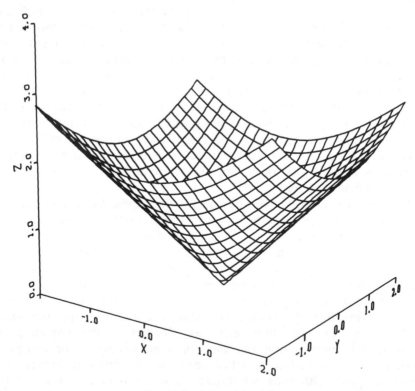

FIG. 11.1   Cone $\varphi(p) = |p|$.

Before proceeding with the technical details, some remarks concerning notation are in order. For simplicity, a point with coordinates (x, y) (in the xy-plane) will be denoted by the single letter p, while fixed points at which elevations etc. are supplied will be denoted $p_1, p_2, \ldots, p_N$, each being in fact a number pair: $p_i = (x_i, y_i)$. When we write $f(p)$, we mean, of course, $f(x, y)$. The given information about f consists of its N values $f_i = f(p_i) = f(x_i, y_i)$ for $i = 1, 2, \ldots, N$. Further, the analogue of "absolute value" is $|p|$, which is defined to be the (Euclidean) distance of p from the origin. Thus,

$$|p| = \sqrt{x^2 + y^2}, \qquad p = (x, y).$$

Also, $|p_i - p_j|$ is the distance between p_i and p_j, so that

$$|p_i - p_j| = \sqrt{(x_i - x_j)^2 + (y_i - y_j)^2} = |p_j - p_i|.$$

We can now express the rotational symmetry requirement by demanding that the values of the basic function depend only on the absolute value of p. Perhaps the simplest useful function of this form is

$$\varphi(p) = |p|. \tag{11.1}$$

In this case the Grammian (cf. Eq. 3.1) associated with the basis of translates $\varphi_i(p) = |p - p_i|$ is the $N \times N$ matrix

$$V = [\varphi_i(p_j)] = \left[|p_i - p_j| \right]$$

$$= \begin{bmatrix} 0 & |p_1 - p_2| & \cdots & |p_1 - p_N| \\ |p_2 - p_1| & 0 & & |p_2 - p_N| \\ \vdots & & \ddots & \vdots \\ |p_N - p_1| & \cdots & & 0 \end{bmatrix}. \tag{11.2}$$

When V is invertible, the coefficients a_i in the interpolant $\Sigma a_i \varphi(p - p_i)$ are given by

$$\mathbf{a} = V^{-1}\mathbf{f}, \tag{11.3}$$

as in Chapter 5. Of course, $\varphi(p - p_i) = |p - p_i| = \sqrt{(x - x_i)^2 + (y - y_i)^2}$ here.

We should inquire as to the nature of the interpolating surface. First, the graph of $z = \varphi(p) = \sqrt{x^2 + y^2}$ is a right-circular cone with its vertex at the origin, opening upward. The graph of the translate $\varphi(p - p_i)$ has its vertex at p_i. Consequently, the interpolant, being a linear combination of cones, may have sharp points at the points p_i. To get more insight into the shape of the interpolant, consider a cross-section of the surface by a vertical

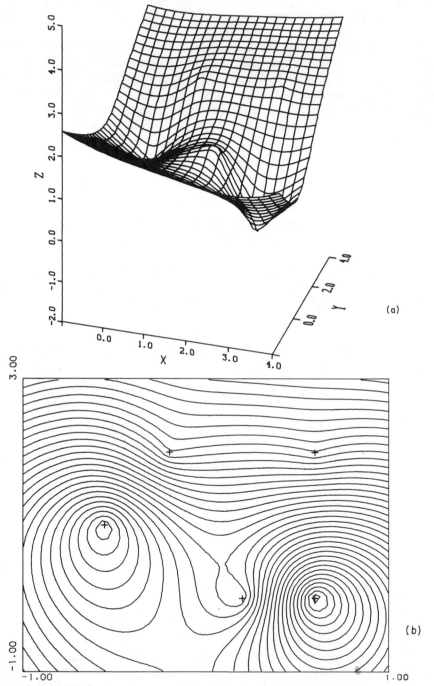

FIG. 11.2 Interpolation with conical basic function. (a) Perspective view and (b) contour map.

plane. This cross-section is a linear combination of the sections of the cones. It is well known that a cone of the form considered here yields a branch of an hyperbola when cut by the plane in question (see Fig. 11.1). If the plane happens to pass through the vertex, the hyperbola degenerates to a V-shaped curve like $|x|$. Consequently, the section is smooth only when it does not pass through a data point. Since each hyperbola has asymptotes which are linear, so does the section in the sense that, far from the data, the cross-section is asymptotic to a straight line.

These properties are illustrated in Fig. 11.2. A surface is constructed which takes assigned heights z at five points (x, y) as follows:

x	2	3	3	1	0
y	0	0	2	2	1
z	2	0	3	3	1

The lack of smoothness at the data points and the linear trend at a distance from these points are apparent.

This interpolation process will not reproduce a plane. To obtain a related interpolant which will do so, one may form a Boolean sum (ref. Section 5.3) of this projector with a projector onto $\mathcal{P}_1$, the vector space of first-degree polynomials in x and y. We consider this in subsequent sections.

We have dwelt on this particular interpolant for several reasons. One is that the *Hardy multiquadric method* (see Hardy, 1971) is very closely related and can give good results. In this method, the cone vertex is rounded off by using the modified basic function

$$\varphi(p) = \sqrt{|p|^2 + r^2},\tag{11.4}$$

where r is a parameter. This function, whose graph is the upper portion of an hyperboloid (Fig. 11.3), is actually very smooth, and a surface derived from it should perhaps not be called a surface spline.

Another reason for considering the interpolant based on $|p|$, is that it has frequently been employed in the kriging method of estimation of function values at points other than the data points. We discuss this in Section 11.6.

The data used for Fig. 11.2 is now interpolated using the basic function of Eq. (11.4) with $r = 0.6$. The result is illustrated in Fig. 11.4. We see that the desired smoothness of the interpolant is attained.

In Figs. 11.5–11.8 we show the results of some experiments using multiquadric interpolation on the standard example of Section 7.7. In Figs. 11.5 and 11.6 surfaces are fitted at 40 points chosen at random. Figure 11.5 indicates the dramatic effect of varying the parameter r and suggests that some thought is necessary in determining an appropriate value. The curious "noisy" surface generated when $r = 0.6$ is, we conjecture, due to the ill-conditioning of the matrix V in Eq. (11.3), which is likely to arise when r is large compared to the spacing of the data points. [See Franke (1979, 1982) for more examples and discussion.]

In Figs. 11.7 and 11.8 we set the parameter r equal to the value 0.1 and show results for surfaces at three different randomly selected sets of data points. The set of Figs. 11.7b and 11.8b is that used in Figs. 11.5 and 11.6.

We conclude this section with examples of functions φ that can be used for interpolation. At this time there is no guarantee that all of the associated Grammians are invertible for all choices of distinct points p_i. Some of these functions have subtle properties to which we return later, in Section 11.5.

EXAMPLE 1 $\varphi(p) = |p|^3$. This analogue of the basis used for cubic splines is smoother than $|p|$, but does not have a smooth second partial derivative.

EXAMPLE 2 $\varphi(p) = |p|^{2k+1}$, $k = 0, 1, 2, \ldots$. As k increases, so does the smoothness. Even powers will not work, because $|p|^{2k}$ is just a polynomial of degree $2k$, and its translates are not linearly independent. Certain linear combinations of these functions are used in kriging.

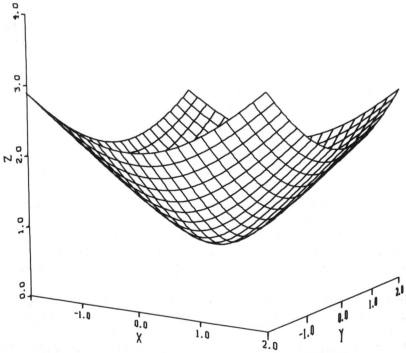

FIG. 11.3   Hardy multiquadric function $\varphi(p) = \sqrt{|p|^2 + r^2}$.

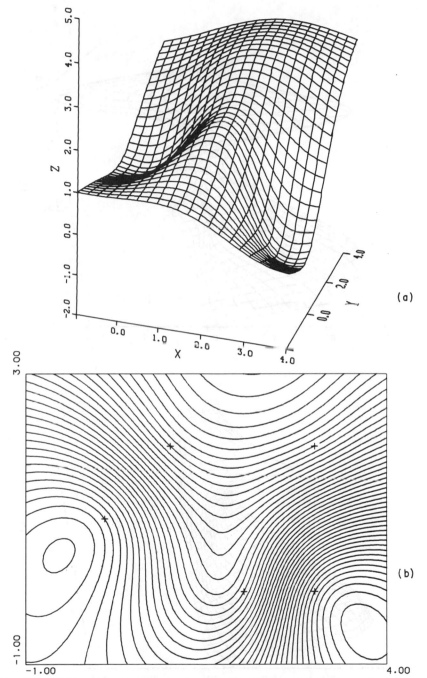

FIG. 11.4   Multiquadric interpolant to the data of Fig. 11.2, $r = 0.6$. (a) Perspective view and (b) contour map.

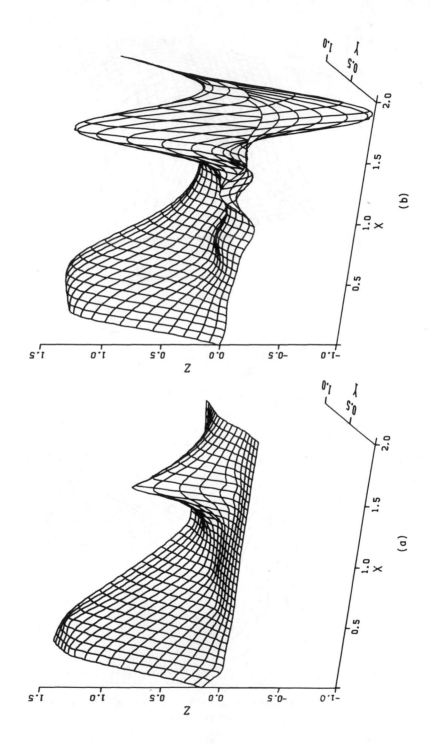

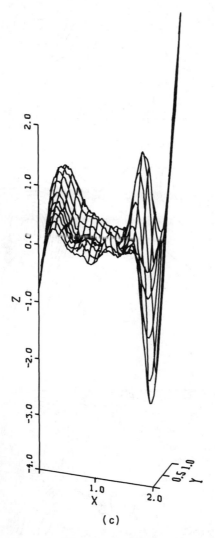

(c)

FIG. 11.5   Multiquadric interpolants; three values of *r*. (a) *r* = 0, (b) *r* = 0.2, and (c) *r* = 0.6.

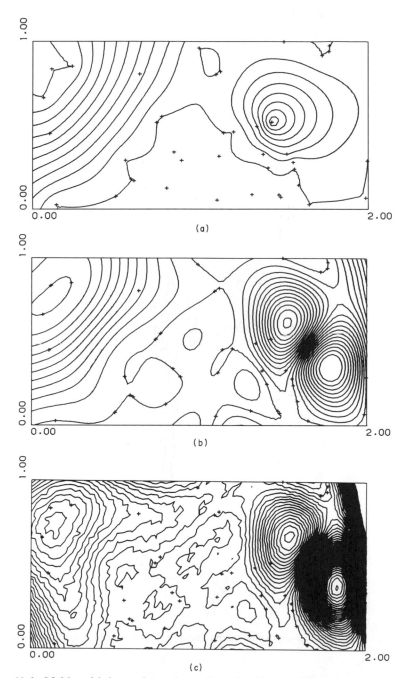

FIG. 11.6   Multiquadric interpolants; three values of r. (a) $r = 0$, (b) $r = 0.2$ and (c) $r = 0.6$.

EXAMPLE 3 $\varphi(p) = |p|^2 \ln|p|$. When used in a certain Boolean sum, this yields the so-called thin-plate spline, a minimum-energy analogue of the natural cubic spline. It does not have a smooth second derivative.

EXAMPLE 4 $\varphi(p) = e^{-\alpha|p|^2}$, $\alpha > 0$. This so-called rotated Gaussian is very smooth. The interpolant is sensitive to the value of the parameter α.

All of these interpolants are *global* in that every data point enters into the Grammian. Consequently, they may be expensive and time-consuming to compute. By contrast, the finite-element methods are local.

11.3 Interpolation with Boolean sums

All of the interpolants of Section 11.2 can be augmented by using them as the interpolating projector P in Boolean sums, exactly as in Section 5.5. For this we must select a projector Q onto the space of polynomials that one wishes to preserve. The main purpose of this section is to indicate how the constructions used in curve fittings are generalized to give constructions for projectors Q onto spaces of bivariate polynomials.

As a first example, for the Boolean sum interpolant to preserve planar data by returning a planar interpolant, we may choose any three non-collinear points and define Qf as the plane interpolant to f at these points. This plane is uniquely defined and can be expressed in terms of three cardinal functions and the three values of f at the chosen points (cf. Section 7.4).

More generally, one may choose to define Q in terms of a weighted least squares polynomial approximant to f, involving all the f-values and a weight matrix that has still to be chosen. Suppose that the approximating function is to be a bivariate polynomial from the space $\mathcal{P}_m$ (see Section 7.1). The procedure is the two-dimensional analogue of that discussed in Section 5.3 and follows the format of Section 7.7 with a modification because of the use of weights. Accordingly, Eq. (7.14) is modified to

$$V^T W V \mathbf{a} = V^T W \mathbf{f}, \qquad (11.5)$$

the element w_{ij} of the matrix W being the weight assigned to the contribution of $[p(x_i, y_i) - f_i][p(x_j, y_j) - f_j]$ to the error $E(p)$. The freedom of choice that we have for the matrix W will be a vital point in the construction of Section 11.4.

If W is positive-definite and V has full rank (ref. Section 7.7), then Eq. (11.5) yields coefficients $\mathbf{a}$ so that the corresponding polynomial is an optimal approximant of the data with respect to W. Even if W is not

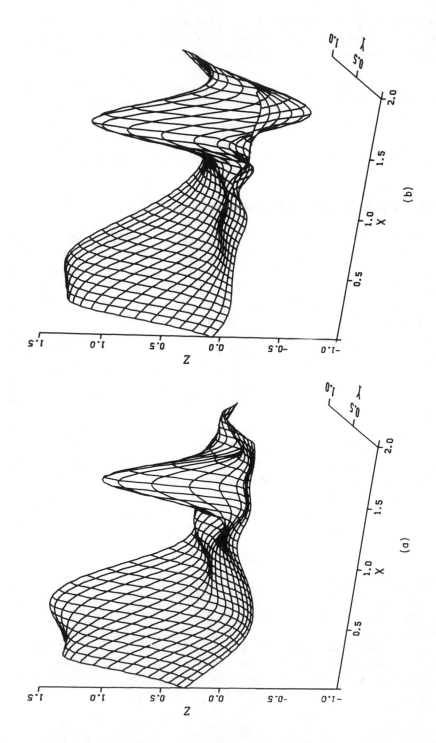

(a)

(b)

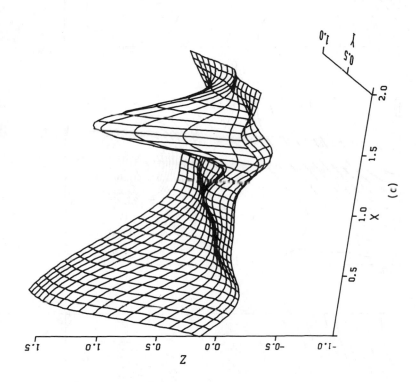

FIG. 11.7 Multiquadric interpolants; three data sets; $r = 0.1$. (a) First data set, (b) second data set and (c) third data set.

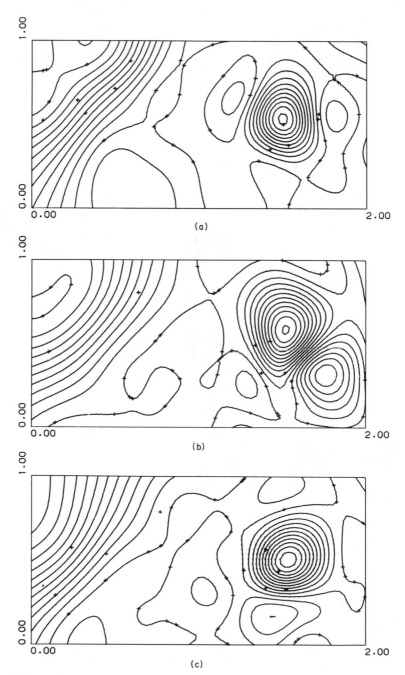

Fig. 11.8 Multiquadric interpolants; three data sets. (a) First data set, (b) second data set and (c) third data set.

positive-definite, provided $V^T W V$ is non-singular, a projector Q_W onto the space $\mathscr{P}_m$ of polynomials is defined by the coefficients **a**. The table in Section 7.2 shows how long the vector **a** will have to be for some values of m, the degree of the approximating polynomial.

Regardless of how the projector Q_W is constructed, its final form will be

$$
Q_W f = [b_1, \quad b_2, \quad \ldots, \quad b_M] \mathfrak{A} \begin{bmatrix} f_1 \\ \vdots \\ f_N \end{bmatrix}, \tag{11.6}
$$

where $b_1, \ldots, b_M$ are the basis functions for $\mathscr{P}_m$ (see the table in Section 7.1) and $\mathfrak{A}$ is an $M \times N$ matrix.

In Section 11.4, next, we discuss one way of optimizing some Boolean sum interpolants by an appropriate choice of W. In the remainder of this section we show that this formulation does, indeed, include the example mentioned in the first paragraph of this section.

The formulation of the projector Q_W in terms of a weighted least squares approximation does not prevent it from actually being an interpolating projector at a subset of the N given points, as we shall now illustrate. For simplicity, we suppose that $Q_W f$ is to be a plane (i.e, $m = 1$) interpolant to f at the non-collinear points p_1, p_2, p_3. Take for W the $N \times N$ matrix

$$
\begin{bmatrix}
1 & 0 & 0 & \vdots & 0 \\
0 & 1 & 0 & \vdots & 0 \\
0 & 0 & 1 & \vdots & 0 \\
\hdashline
0 & 0 & 0 & \vdots & 0
\end{bmatrix}, \tag{11.7}
$$

in which the top left corner is the 3×3 identity matrix, and the rest consists of zeros. Now we compute the coefficient matrices from Eq. (11.5). It will be found that $V^T W V$ collapses to

$$
\begin{bmatrix}
1 & 1 & 1 \\
x_1 & x_2 & x_3 \\
y_1 & y_2 & y_3
\end{bmatrix}
\begin{bmatrix}
1 & x_1 & y_1 \\
1 & x_2 & y_2 \\
1 & x_3 & y_3
\end{bmatrix},
$$

whereas $V^T W$ becomes

$$
\begin{bmatrix}
1 & 1 & 1 & 0 & \ldots & 0 \\
x_1 & x_2 & x_3 & 0 & \ldots & 0 \\
y_1 & y_2 & y_3 & 0 & \ldots & 0
\end{bmatrix}.
$$

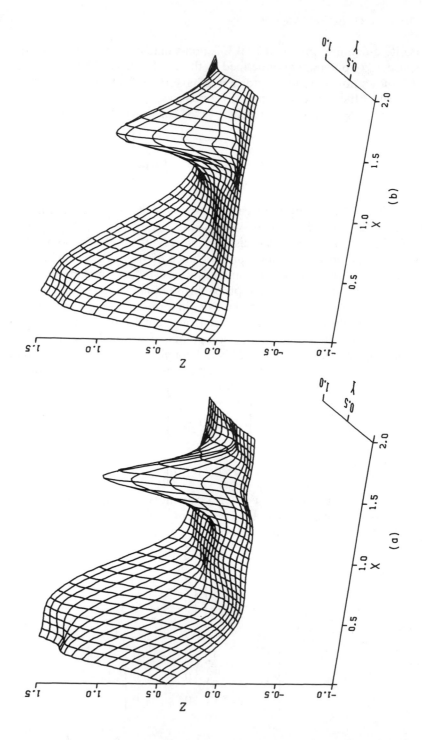

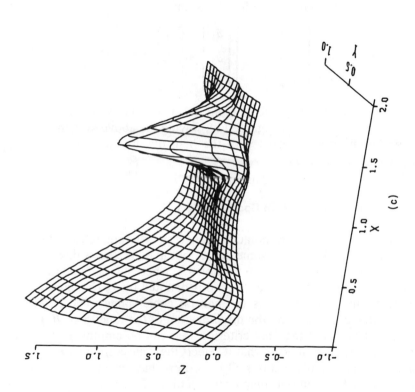

(c)

FIG. 11.9   Thin plate spline interpolants; three data sets. (a) First data set, (b) second data set and (c) third data set.

Consequently, $V^T W \mathbf{f}$ is the same as

$$
\begin{bmatrix} 1 & 1 & 1 \\ x_1 & x_2 & x_3 \\ y_1 & y_2 & y_3 \end{bmatrix} \begin{bmatrix} f_1 \\ f_2 \\ f_3 \end{bmatrix}.
$$

The small 3×3 Vandermonde matrix is non-singular in view of the non-collinearity of p_1, p_2, p_3 and can now be cancelled from both sides of the equation. Thus, we obtain the reduced system

$$
\begin{bmatrix} 1 & x_1 & y_1 \\ 1 & x_2 & y_2 \\ 1 & x_3 & y_3 \end{bmatrix} \begin{bmatrix} a_1 \\ a_2 \\ a_3 \end{bmatrix} = \begin{bmatrix} f_1 \\ f_2 \\ f_3 \end{bmatrix},
$$

which simply says that the plane $a_1 + a_2 x + a_3 y$ has elevations f_1, f_2, f_3 at p_1, p_2, p_3. This plane can be obtained by solving the above system for a_1, a_2, a_3, or by any other means; with the help of cardinal functions, for example (see Section 7.4).

11.4 Optimal interpolation with Boolean sums

In the construction of any interpolation scheme, it is natural to ask whether the resulting interpolant is, in some sense, optimal. We have discussed a univariate example of optimality in Section 4.4, where it is claimed that the best interpolant from the class of functions denoted by $\mathcal{F}_2$ (having square integrable second derivative), is the natural cubic spline. The criterion "best" is in that case based on the minimization of the functional $J(s)$ [see Eq. (4.19)]. This is not the only criterion that can be employed; for our present purposes it is significant that it selects the *natural spline function* as best among some class of functions. Other criteria have been devised which optimize not the function but some property of the function *at a point*. We shall elaborate on this next in Section 11.5.

In this section we record the existence of a bivariate analogue of the natural cubic spline which is also optimal in an appropriate sense. This is the so-called thin-plate spline, which, while interpolating, also minimizes a functional $J(s)$ analogous to that defined by Eq. (4.19). This functional is

$$
J(s) = \int_{\mathbf{R}^2} \int \left\{ \left[\frac{\partial^2 s}{\partial x^2} \right]^2 + 2 \left[\frac{\partial^2 s}{\partial x \, \partial y} \right]^2 + \left[\frac{\partial^2 s}{\partial y^2} \right]^2 \right\} dx \, dy. \qquad (11.8)
$$

In contrast to the situation in equation (4.19), the integration here is to be carried out over the whole plane, hence $J(s)$ will exist only for rather special functions whose second derivatives are close to zero far away from the origin, i.e. functions which flatten out rapidly. The functional $J(s)$ is an approximation to the energy of a thin plate of infinite extent which deforms only by bending. It turns out that the interpolant s which minimizes $J(s)$ can be expressed as a Boolean sum:

Theorem 11.4.1 *Subject to the invertibility of the matrices V and $B^T V^{-1} B$ (see below), the thin plate spline s is the Boolean sum $s = (P \oplus Q_W)f$, where P is the interpolating projector formed from the translates of the basic function $\varphi(p) = |p|^2 \ln|p|$, and Q_W is the projector onto $\mathcal{P}_1$ defined by the weighted least squares process [Eq. (11.5)], with weight matrix W equal to the inverse of the Vandermondian of the translates of the basic function.*

In symbols, Theorem 11.4.1 states that

$$s - [\varphi(p - p_1), \ldots, \varphi(p - p_N)] V^{-1} \begin{bmatrix} f_1^* \\ \vdots \\ f_N^* \end{bmatrix}$$

$$+ [1, x, y](B^T V^{-1} B)^{-1}(B^T V^{-1}) \begin{bmatrix} f_1 \\ \vdots \\ f_N \end{bmatrix},$$

where V is the (symmetric) Vandermondian of the translates of $\varphi(p)$; the vector $[f_1^*, \ldots, f_N^*]^T$ is equal to $[f_1, \ldots, f_N]^T$ reduced by the values of the projector $Q_W f$:

$$\mathbf{f}^* = \mathbf{f} - B(B^T V^{-1} B)^{-1} B^T V^{-1} \mathbf{f},$$

and

$$B^T = \begin{bmatrix} 1 & 1 & \cdots & 1 \\ x_1 & x_2 & \cdots & x_N \\ y_1 & y_2 & \cdots & y_N \end{bmatrix}.$$

This formulation of s is, except for small details, just like that of the natural spline, as in Theorem 5.4.1 and its corollary 5.4.2.

Computationally, this surface spline is an expensive function. In the univariate case it is possible to reduce computational effort by exploiting the fact that the natural spline can be computed in a piecewise fashion once the simple tridiagonal system [Eq. (4.6)] has been solved. When we move

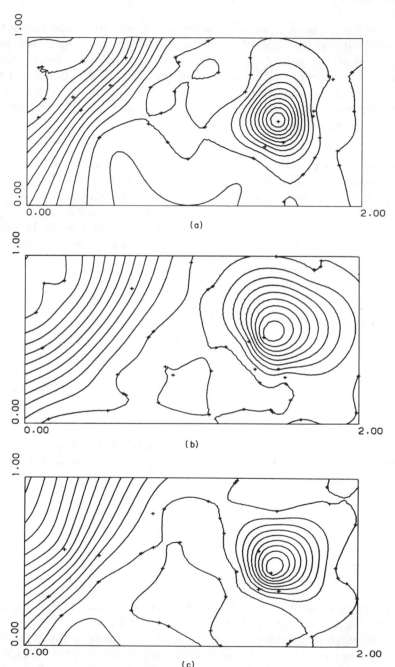

FIG. 11.10 Thin plate spline interpolants; three data sets. (a) First data set, (b) second data set and (c) third data set.

to the bivariate case, the piecewise approach separates from the functionally optimal one to become a finite-element method.

Of course, the basic function here is intimately related to the functional $J(s)$ whose minimization has an obvious physical interpretation. A number of similar functionals and associated basic functions appear in the technically advanced work of Meinguet (1978, 1979).

Figs. 11.9 and 11.10 show the appearance of thin plate spline interpolants to the standard surface. The data sets are simply those of Figs. 11.7 and 11.8.

In Section 11.5 we next present a set of interpolation methods which are based on a more elementary optimization criterion.

11.5 Optimization at the points of interpolation

We shall now describe a process of optimization of the interpolant at a point. Computationally the result is identical with the probabilistic kriging method, and in some cases it is also functionally optimal in the sense of the thin plate spline.

As before, let f_i, $i = 1, \ldots, N$ denote the given values of $f(x, y)$ at the distinct points $p_i = (x_i, y_i)$, $i = 1, \ldots, N$. Then, whichever interpolating projector P is constructed, the interpolant has a representation in terms of the cardinal functions λ_i associated with P:

$$(Pf)(p) = \sum_{i=1}^{N} \lambda_i(p)f_i, \qquad p = (x, y). \tag{11.9}$$

When p is fixed, the value of Pf at p [denoted above by $(Pf)(p)$] is just a linear combination of the known function-values. We impose the condition that P preserve polynomials of degree m, and that m is much smaller than the number N of data points. This means that $Pf = f$ when $f(p) = 1, x, y, \ldots$, up to terms of degree m in x and y. The table in Section 7.2 shows what the polynomial basis functions are for some small values of m. If we denote the dimension of $\mathcal{P}_m$ by M, then we may conveniently write a basis for $\mathcal{P}_m$ as consisting of the functions $b_1, \ldots, b_M$, with $b_1 = 1$, $b_2 = x$, etc., and our requirement is that

$$Pb_j = b_j, \qquad j = 1, \ldots, M, \tag{11.10}$$

or, in view of Eq. (11.9),

$$\sum_{i=1}^{N} \lambda_i(p)b_j(p_i) = b_j(p), \qquad j = 1, \ldots, M. \tag{11.11}$$

Now, with p still fixed, the error at p committed by interpolation is

$$E(p) = f(p) - \sum_{i=1}^{N} \lambda_i(p)f_i. \tag{11.12}$$

This can be written as the (dot) product of two vectors:

$$E(p) = -[f(p), f_1, \ldots, f_N][-1, \lambda_1(p), \ldots, \lambda_N(p)]^T. \tag{11.13}$$

As well, Eq. (11.11) can be written in the vector-matrix form

$$\begin{bmatrix} b_1(p) & b_1(p_1) & \cdots & b_1(p_N) \\ b_2(p) & b_2(p_1) & \cdots & b_2(p_N) \\ & & \vdots & \\ b_M(p) & b_M(p_1) & \cdots & b_M(p_N) \end{bmatrix} \begin{bmatrix} -1 \\ \lambda_1(p) \\ \vdots \\ \lambda_N(p) \end{bmatrix} = \begin{bmatrix} 0 \\ \vdots \\ 0 \end{bmatrix}, \tag{11.14}$$

which we shall abbreviate using partitioned matrices as

$$\left[\mathbf{b}^T(p) \,\middle|\, B^T \right] \begin{bmatrix} -1 \\ \hline \boldsymbol{\lambda} \end{bmatrix} = \begin{bmatrix} 0 \\ \hline 0 \end{bmatrix}. \tag{11.14a}$$

Equation (11.14a) is a constraint on the permissible values of $\lambda_1(p), \ldots,$ $\lambda_N(p)$. At this point it may be useful to review the concept of *vector norm*. This is a way of attributing length to a vector. In the simplest case, let $\mathbf{x}^T = [x_1, \ldots, x_M]$ be a vector with M components. Its *Euclidean norm* or length is given by the familiar expression

$$|\mathbf{x}| = (x_1^2 + x_2^2 + \ldots + x_M^2)^{1/2}.$$

An alternative form for this is

$$|\mathbf{x}| = (\mathbf{x}^T\mathbf{x})^{1/2}.$$

A different length will emerge if we modify this expression by choosing a positive-definite $M \times M$ matrix W and define the W-norm (or W-length) of $\mathbf{x}$ by

$$|\mathbf{x}|_W = (\mathbf{x}^T W \mathbf{x})^{1/2}. \tag{11.15}$$

The positive-definiteness of W guarantees that $|\mathbf{x}|_W \geqslant 0$. Now, it is possible to show that the magnitude of $E(p)$ in Eq. (11.13) can be bounded as follows. For any positive-definite $(N + 1) \times (N + 1)$ matrix W,

$$|E(p)| \leqslant C|[-1, \lambda_1(p), \ldots, \lambda_N(p)]|_W, \tag{11.16}$$

where the constant C depends on W and the values $f(p), f(p_1), \ldots, f(p_N)$.

This is a consequence of the well known Cauchy–Schwarz inequality [see Lancaster and Tismenetsky (1985)]. In fact, since the λ_i's are constrained by Eq. (11.14), W need not be positive-definite for all vectors of length $(N + 1)$—only for those satisfying (11.14). Such a matrix has been called *conditionally positive-definite* of *order $m + 1$*, m being the maximum degree of the polynomials involved in Eq. (11.10). For our purposes we shall insist that W be symmetric.

We come now to the optimization of the quantity $\Sigma_{i=1}^{N} \lambda_i(p) f_i$. Ideally, we would like to minimize the error $E(p)$, but since $f(p)$ is unknown, this is not possible. Instead, we choose a suitable W and minimize the upper bound on $|E(p)|$, which appears in Eq. (11.16):

Definition 11.5.1   *Let W be an $(N + 1) \times (N + 1)$ matrix which is conditionally positive-definite of order $m + 1$. The **optimal approximation to** $f(p)$ **with respect to** W has the form $\Sigma_{i=1}^{N} \lambda_i f_i$, where the λ_i's are chosen to minimize*

$$\|[-1, \lambda_1, \ldots, \lambda_N]\|_W \tag{11.17}$$

subject to the constraints of Eq. (11.14). (For simplicity of notation, the dependence of the λ_i's on p has not been indicated here.)

This minimization problem can be handled very easily by the method of Lagrange multipliers. However, a few remarks concerning the result of the optimization can already be made. First, in order to obtain an interpolant, W will have to depend on the variable p and the given points p_i in a special way. A second observation is that although this optimization is done at a point p, if its numerical coordinates are left as variables, then the result will be a function of x and y.

Third, the resulting interpolant is not optimal in any universal sense. For each proper choice of W the result is optimal in that it minimizes the corresponding W-norm of the vector involved in the bound of Eq. (11.16). It is a different question whether the chosen W results in an interpolant which is optimal for the data.

We now indicate the result of the minimization. This is carried out with the aid of Lagrange's multipliers $\mu_1, \mu_2, \ldots, \mu_M$, which are equal in number to the number of equations in the constraints of Eq. (11.10). The technique will be found in almost any advanced calculus book. One obtains the following result concerning the optimal values of $\lambda_1, \ldots, \lambda_N$ and the associated multipliers $\mu_1, \ldots, \mu_M$ (for which we have no further use).

Theorem 11.5.1 *The vectors* $\boldsymbol{\lambda}$ *and* $\boldsymbol{\mu}$ *(or equivalently the numbers* $\lambda_1, \ldots, \lambda_N, \mu_1, \ldots, \mu_M$*) satisfy the system of equations*

$$\left[\begin{array}{c|c} V & B \\ \hline B^T & 0 \end{array}\right] \left[\begin{array}{c} \boldsymbol{\lambda} \\ \hline \boldsymbol{\mu} \end{array}\right] = \left[\begin{array}{c} \mathbf{v} \\ \hline \mathbf{b}(p) \end{array}\right], \tag{11.18}$$

in which B *and* $\mathbf{b}(p)$ *are as in Eqs.* (11.14) *and* (11.14a), *and* V *and* v *are portions of the symmetric matrix* W *obtained by partitioning it as indicated:*

$$W = \left[\begin{array}{c|c} u & \mathbf{v}^T \\ \hline & \\ \mathbf{v} & V \end{array}\right]. \tag{11.19}$$

By using techniques appropriate to partitioned matrices, the required solution $\boldsymbol{\lambda}$ can be displayed explicitly. One finds that there are two matrices P and Q such that

$$\boldsymbol{\lambda} = P\mathbf{v} + Q\mathbf{b}(p). \tag{11.20}$$

Here, P and Q have a rather complicated form, but will be seen to be just like the expressions appearing in the Boolean sum of Theorem 11.4.1 and earlier related theorems. Thus, provided the indicated matrix inversions are possible,

$$P = V^{-1} - [V^{-1}B][B^TV^{-1}B]^{-1}[B^TV^{-1}], \quad Q = [V^{-1}B][B^TV^{-1}B]^{-1}, \tag{11.21}$$

and $P^T = V^{-1} - V^{-1}B[B^TV^{-1}B]^{-1}B^TV^{-1}$ and $Q^T = [B^TV^{-1}B]^{-1}B^TV^{-1}$. Consequently, the estimate $f(p)$ is given by

$$f(p) = \boldsymbol{\lambda}^T\mathbf{f} = \mathbf{v}^TP^T\mathbf{f} + \mathbf{b}^T(p)Q^T\mathbf{f}. \tag{11.22}$$

We can see here that the second term is just like a weighted least squares approximation to the data, using a weight matrix V^{-1}. Since V is not necessarily positive-definite, this is not a proper least squares approximation. Nevertheless, it is a projector onto $\mathscr{P}_m$, since $\mathbf{b}(p)$ is a vector of basis polynomials for $\mathscr{P}_m$ (see also the comments in Section 11.3). Denote the projector by $\mathscr{Q}$.

Concerning the second term, we see that $P\mathbf{v}$ contains $V^{-1}\mathbf{v}$, an expression reminiscent of the construction of a cardinal basis from the (possibly non-cardinal) basis functions which make up the vector $\mathbf{v}$. This shows us what can be done to make this whole process interpolate. We choose W so that apart from the top left element the first column and first row consist of some basis functions $\varphi_1, \ldots, \varphi_N$, and make the large block V in W [see

Eq. (11.19)] equal to the Vandermondian (or Grammian) of the φ_i's. A simple way to achieve this is to choose for each $\varphi_i(p)$ the translate to p_i of a basic function $\varphi(p)$, and to let W be the matrix

$$
W = \begin{bmatrix}
\varphi(p-p) & \varphi(p_1-p) & \cdots & \varphi(p_N-p) \\
\varphi(p-p_1) & \varphi(p_1-p_1) & \cdots & \varphi(p_N-p_1) \\
\vdots & \vdots & \vdots & \vdots \\
\varphi(p-p_N) & \varphi(p_1-p_N) & \cdots & \varphi(p_N-p_N)
\end{bmatrix}. \qquad (11.23)
$$

Then the requirements for interpolation will be fulfilled, and $V^{-1}v$ is a cardinal basis vector for an interpolating projector $\mathcal{P}$. Then Eq. (11.22) is just a Boolean sum $\mathcal{P} \oplus \mathcal{Q}$ of the two projectors exactly as in Section 11.4.

It remains for us to comment on the choice of φ. This choice has to be such that W will be conditionally positive-definite of order $m + 1$. The theory of such functions is sophisticated and difficult. It also only guarantees *semi*-definiteness. Functions which have been used are:

(1) $\varphi(p) = e^{-\alpha|p|^2}$, which yields a matrix conditionally positive-semi-definite of any order;
(2) $(-1)^{k+1}|p|^{2k+1}$, $0 \leqslant k \leqslant m$, W positive-semi-definite of order $m + 1$; and
(3) $|p|^2 \ln|p|$, W positive-semi-definite of order $m + 1 = 2$.

Interpolants using the first function will be very smooth, whereas those using the second will have some discontinuous derivatives. In particular, the choice $k = 0$ yields the surface which was discussed at some length in Section 11.2. In a one-dimensional version of this, the choice $k = 1$ and $m = 2$ in item 2 yields natural cubic splines. The last choice is the one which generates the thin-plate spline.

11.6 Kriging

In recent years, a surface-fitting technique known as "kriging" has been used with increasing frequency, especially in geostatistics. Although this book is primarily concerned with deterministic methods, it will be useful to discuss the most elementary forms of kriging and to show that these simple forms are identical with the Boolean sum processes of Section 11.5. What distinguishes kriging from our approach is that, in the hands of its practitioners, attempts are made to

(1) localize the computation by excluding distant points from the calculation of the interpolant at any fixed point p, and

(2) analyse the data in order to assist in the choice of an appropriate conditionally-positive-function.

A context in which kriging is used is that of producing contour maps of surfaces derived from irregularly scattered points in the plane. There are, of course, other techniques that can be applied here, as discussed in Chapters 9 and 10. Because some contouring processes cannot cope directly with data points scattered irregularly in the plane, kriging is used to estimate values at the points of a regular rectangular grid. The original data is then discarded, and the new data contoured. However, the contouring process itself contains, implicitly at least, another interpolation process which may not even be known to the user if the contouring method is part of a computer package. The result is that the contour map so generated is neither a map of an interpolant of the original data nor of the kriging interpolant. In this way, the actual features of the kriging interpolant may be obscured. This is especially "fortunate" when the latter is rough, as is the case when the positive-definite function $\varphi(p) = -|p|$ is used. The contouring process may then serve to "smooth" the surface generated by kriging. This situation can be described as the composition of two projectors that do not sample the data at the same points (cf. Section 10.5).

Let us sketch some of the underlying ideas of kriging. We assume that the data is a sample from a random function $f(p)$, which is the sum of a "slowly" (space) varying random polynomial $d(p)$ of degree m, called the drift, and a "rapidly" (space) varying random component $r(p)$, which is assumed to have zero mean or expected value $E(r) = 0$, and which is responsible for the noise-like nature of $f(p)$. Thus,

$$f(p) = d(p) + r(p), \qquad E(r) = 0. \tag{11.24}$$

One assumes further that the covariance structure of $r(p)$ can be obtained and that the covariance between values of $r(p)$ at points p and q depends only on the distance between p and q. Then [see Eq. (11.1)],

$$\text{cov}[f(p), f(q)] = \Phi(|p - q|). \tag{11.25}$$

Now we are to estimate $f(p)$ by the linear combination $\sum_{i=1}^{N} \lambda_i f(p_i)$ in such a way that the variance of the error $f(p) - \sum_{i=1}^{N} \lambda_i f(p_i)$ is minimized. At the same time, the λ_i's should be defined in such a way that the variance is independent of the drift $m(p)$, for nothing is assumed about the statistical properties of $m(p)$. This is achieved by applying to the λ_i's the constraints of Eq. (11.11) or Eq. (11.14). The result of this optimization is a vector $\boldsymbol{\lambda}$ given by (11.20) provided we choose

$$
W = \begin{bmatrix}
\Phi(|p - p|) & \Phi(|p - p_1|) & \cdots & \Phi(|p - p_N|) \\
\Phi(|p_1 - p|) & \Phi(|p_1 - p_1|) & \cdots & \Phi(|p_1 - p_N|) \\
\vdots & \vdots & & \vdots \\
\Phi(|p_N - p|) & \Phi(|p_N - p_1|) & \cdots & \Phi(|p_N - p_N|)
\end{bmatrix}. \quad (11.26)
$$

Of course, this covariance matrix W is identical in form with that in Eq. (11.23), since $\varphi(p) = \Phi(|p|)$.

The main thrust of kriging is to choose Φ in a way that is consistent with the data. This is the source of the claim made by some, that kriging is an optimal process of interpolation.

For an introduction to kriging in an entirely statistical setting, we refer the reader to the work of Olea (1975) and Huijbregts and Matheron (1970). It will be found there that the function $\Phi(|p|)$ need not be an ordinary covariance, decaying to zero with increasing values of $|p|$; rather, semi-variograms and generalized covariances related to the conditionally positive-definite functions of Section 11.5 come into play.

References

Ahlberg, J. H. (ed.) (1970). "Spline Approximation and Computer-Aided Design." Academic Press, London and Orlando.

Ahlberg, J. H., Nilson, E. N. and Walsh, J. L. (1967). "The Theory of Splines and their Applications." Academic Press, London and Orlando.

Akima, H. (1970). A new method of interpolation and smooth curve fitting based on local procedures. *J. Assoc. Comp. Mach.* **17**, 589–602.

Akima, H. (1972). Interpolation and smooth curve fitting based on local procedures. *Comm. ACM* **15**, 914–918.

Alfeld, P. and Barnhill, R. E. (1977). A transfinite $\mathscr{C}^2$ interpolant over triangles. *Rocky Mountain J. Math.* **14**, 17–39.

Barnhill, R. E. (1977). Representation and approximation of surfaces. In "Mathematical Software", Vol. III (J. R. Rice, ed.), pp. 68–119. Academic Press, London and Orlando.

Barnhill, R. E., Dube, R. P. and Little, F. F. (1983). Properties of Shepard's surfaces. *Rocky Mountain J. Math.* **13**, 365–382.

Bengtsson, B. E. and Nordbeck, S. (1964). Construction of isarithms and isarithmic maps by computers. *B.I.T.* **4**, 87–105.

Bhattacharyya, B. K. (1969). Bicubic spline interpolation as a method for treatment of potential field data. *Geophysics* **34**, 402–423.

Birkhoff, G. (1967). Local spline approximation by moments. *J. Math. and Mech.* **16**, 987–990.

Boehm, W., Farin, G. and Kahmann, J. (1984). A survey of curve and surface methods in CAGD. *Computer Aided Geometric Design* **1**, 1–60

Bos, L. P. and Šalkauskas, K. (1987). On the Matrix $[|x_i - x_j|^3]$ and the Cubic Spline Continuity Equations. *J. Approximation Theory*, 51, 81–88.

Briggs, I. C. (1974). Machine contouring using minimum curvature. *Geophysics* **39**, 39–48.

Ciarlet, P. (1978). "The Finite Element Method for Elliptic Problems." North Holland, Amsterdam.

Cline, A. K. and Renka, R. L. (1984). A storage-efficient method for construction of a Thiessen triangulation. *Rocky Mountain J. Math.* **14**, 119–139.

Clough, R. W. and Tocher, J. L. (1965). Finite element stiffness matrices for analysis of plates in bending. *Proc. Conf. Matrix Methods in Structural Mechanics*. Wright-Patterson A.F.B., Ohio.

Conte, S. D. and de Boor, C. (1980). "Elementary Numerical Analysis" (3rd ed.). McGraw Hill, New York.

Cox, M. G. (1975). An algorithm for spline interpolation. *J. Inst. Math. Appl.* **15**, 95–108.

Crain, E. R. (1972). Review of gravity and magnetic data processing systems. *Can. Soc. Exp. Geophys. J.* **8**, 54–77.

Crain, I. K. (1970). Computer interpolation and contouring of two-dimensional data: A review. *Geoexploration* **8**, 71–86.

Crain, I. K. and Bhattacharyya, B. K. (1967). The treatment of two-dimensional non-equispaced data with a digital computer. *Geoexploration* **5**, 173–194.

Craven P. and Wahba, G. (1977). Smoothing noisy data with spline functions; estimating the correct degree of smoothing by the method of cross-validation. *Tech. Rep. 445*, Dept. of Stat., Univ. of Wisconsin, Milwaukee.

Davis, J. C. (1973). "Statistics and Data Analysis in Geology." Wiley, New York.

Davis, J. C. and McCullagh, M. J. (eds) (1975). "Display and Analysis of Spatial Data." Wiley, New York.

Davis, P. J. (1975). "Interpolation and Approximation." Dover Publications, New York.

de Boor, C. (1962). Bicubic spline interpolation. *J. Math. Phys.* **41**, 212–218.

de Boor, C. (1972). On calculating with B-splines. *J. Approx. Theory* **6**, 50–62.

de Boor, C. (1977). Package for calculating with B-splines. *SIAM J. Numer. Anal.* **14**, 441–472.

de Boor, C. (1978). "A Practical Guide to Splines." Springer Verlag, New York.

Delfiner, P. (1975). "Linear Estimation of Non-Stationary Spatial Phenomena." Vol. 12. NATO Adv. Study Institute, Rome.

Duchon, J. (1976). Functions-spline et esperances conditionelles de champs gaussiens. *Ann. Sci. Univ. Clermont.* No. 61, 19–27.

Duchon, J. (1977). Splines minimizing rotation-invariant semi-norms in Sobolev spaces. In "Constructive Theory of Functions of Several Variables" (W. Schempp and K. Zeller, eds), Lecture Notes in Mathematics **571**, 85–100. Springer-Verlag, Berlin.

Ferguson, J. and Staley, P. A. (1973). Least squares piecewise cubic curve fitting. *Comm. ACM* **16**, 380–382.

Forsythe, G. E., Malcolm M. A. and Moler, C. B. (1977). "Computer Methods for Mathematical Computations." Prentice Hall, Englewood Cliffs.

Franke, R. (1982). Scattered data interpolation: Tests of some methods. TR NPS-53-79-003. Naval Postgraduate School, Monterey, California.

Franke, R. (1982). Scattered data interpolation: Tests of some methods. *Math. Comput.* **38**, 181–200.

Gelb, A. (ed.) (1974). "Applied Optimal Estimation." M.I.T. Press, Cambridge, Mass.

Gordon, W. J. (1968). Blending function methods of bivariate and multivariate interpolation and approximation. *Res. Rep. GMR-834*. General Motors, Warren, Michigan.

Gordon, W. J. (1969). Spline-blended surface interpolation through curve netvorks. *J. Math. Mech.* **18**, 931–952.

Gordon, W. J. (1969). Free-form surface interpolation through curve networks. *Res. Rep. GMR-921*. General Motors, Warren, Michigan.

Gordon, W. J. and Wixon, J. A. (1978). On Shepard's method of "metric interpolation" to bivariate and multivariate data. *Math. Comput.* **32**, 253–264.

Green, P. J. and Sibson, R. (1978). Computing Dirichlet tesselations in the plane. *Comput. J.* **21**, 168–173.

Greville, T. N. E. (ed.) (1969). "Theory and Applications of Spline Functions." Academic Press, London and Orlando.

Hardy, R. L. (1971). Multiquadric equations of topography and other irregular surfaces. *J. Geophys. Res.* **76**, 1905–1919.

Hardy, R. L. and Gopfert, W. M. (1975). Least squares prediction of gravity anomalies, geoidal undulations and deflections of the vertical with multiquadratic harmonic functions. *Geophys. Res. Letters* **10**, 423–426.

Hayes, J. G. and Halliday, J. (1974). The least squares fitting of cubic spline surfaces to general data sets. *J. Inst. Math. Appl.* **14**, 89–106.

Hemmerle, W. J. (1967). "Statistical Computations on a Digital Computer." Blaisdell Publ. Co., Waltham, Massachusetts.

Huijbregts, C. and Matheron, G. (1970). Universal kriging: An optimal method for estimating and contouring in trend surface analysis. *Can. Inst. Mining Metall.* **12**, 159–169 (C.I.M.M. Special Volume).

Ichida, K., Voshimoto, F. and Kiyoko, T. (1976). Curve fitting by a piecewise cubic polynomial. *Computing* **16**, 329–338.

Karlin, S. (1968). "Total Positivity" (Volume 1). Stanford University Press, Stanford.

Karlin, S. and Studden, W. J. (1966). "Tchebycheff Systems: With Applications in Analysis and Statistics." Interscience, New York.

Lancaster P. and Šalkauskas, K. (1981). Surfaces generated by moving least squares methods. *Math. Comp.* **37**, 141–158.

Lancaster, P. and Tismenetsky, M. (1985). "The Theory of Matrices." Academic Press, London and Orlando.

Lancaster, P. and Watkins, D. S. (1977). Some families of finite elements. *J. Inst. Math. Appl.* **19**, 385–397.

Lawson, C. L. (1977). Software for $\mathscr{C}^1$ surface interpolation. "Mathematical Software", Vol. III, pp. 69–120. Academic Press, London and Orlando.

Lawson, C. L. and Hanson, R. J. (1974). "Solving Least Squares Problems." Prentice Hall, Englewood Cliffs.

Lodwick, G D and Whittle, J. (1970). A technique for automatic contouring of field survey data. *Aust. Comput. J.* **2**, 104–109.

Lyche, T. and Schumaker, L. L. (1973). Computation of smoothing and interpolating natural splines. *SIAM J. Numer. Anal.* **10**, 1027–1038.

Lyche, T. and Schumaker, L. L. (1974). Procedures for computing smoothing and interpolating natural splines. *Comm. ACM* **17**, 463–467.

Malcolm, M. A. (1977). On the computation of non-linear spline functions. *SIAM J. Numer. Anal.* **14**, 254–282.

McLain, D. H. (1974). Drawing contours from arbitrary data points. *Comput. J.* **17**, 318–324.

Meinguet, J. (1978). Multivariate spline interpolation at arbitrary points. In "Polynomial and Spline Approximation" (B. N. Sahney, ed.), pp. 163–190. Reidel, Dordrecht.

Meinguet, J. (1979). Multivariate interpolation at arbitrary points made simple. *Z. Angew. Math. Phys.* **30**, 292–304.

Melkes, F. (1972). Reduced piecewise bivariate Hermite interpolations. *Numer. Math.* **19**, 326–340.

Mitchell, A. R. (1973). Element types and basis functions. In "Numerical Solution of Partial Differential Equations." pp. 107–150. Kjeller, Norway.

Mitchell, A. R. and Wait, R. (1977). "The Finite Element Method in Partial Differential Equations." John Wiley, New York.

Mood, A. M. and Graybill, F. A. (1963). "Introduction to the Theory of Statistics" (2nd ed.). McGraw-Hill, New York.

Morgan, J. and Scott, R. (1975). A nodal basis for $\mathscr{C}^1$ piecewise polynomials of degree $n > 4$ *Math. Comp.* **29**, 736–740.

Olea, R. A. (1975). "Optimum Mapping Techniques using Regionalized Variable Theory." Kansas Geological Survey, Lawrence, Kansas.

Pelto, C. R., Elkins, T. A. and Boy, H. A. (1968). Automatic contouring of irregularly spaced data. *Geophysics* **33**, 424–430.

Percell, P. (1976). On cubic and quartic Clough–Tocher finite elements. *SIAM J. Numer. Anal.* **13**, 100–103.

Powell, M. J. D. and Sabin, M. A. (1977). Piecewise quadratic approximation on triangles. *ACM Trans. Math. Software* **3**, 316–325.

Rhynsburger, D. (1973). Analytic delineation of Thiessen polygons. *Geogr. Anal.* **5**, 133–144.

Ritchie, S. (1978). *Representation of Surfaces by Finite Elements.* M.Sc. Thesis, University of Calgary.

Rogers, C. A. (1964). "Packing and Covering". Cambridge Univ. Press, Cambridge.

Salkauskas, K. (1984). $\mathscr{C}^1$ splines for interpolation of rapidly varying data. *Rocky Mountain J. Math.* **14**, 239–250.

Sard, A. Weintraub, G. (1971).."A Book of Splines." John Wiley, New York.

Schoenberg, I. J. (1964). Spline functions and the problem of graduation. *Proc. Nat. Acad. Sci. (U.S.)* **52**, 947–950.

Schumaker, L. L. (1976). Fitting surfaces to scattered data. In "Approximation Theory", Vol. II, (Lorentz, G. G., Chui, C. K., Schumaker, L. L., eds.), pp. 203–268. Academic Press, London and Orlando.

Shepard, D. (1968). A two-dimensional function for irregularly spaced data. *Proc. ACM Nat. Conf.*, pp. 517–524.

Sibson, R. and Thomson, G. D. (1981). A seamed quadratic element for contouring. *Comp. J.* **24**, 378–382.

Späth, H. (1969). Spline interpolation of degree three. *Comput. J.* **12**, 198.

Späth, H. (1974). "Spline Algorithms for Curves and Surfaces." Utilitas Mathematica, Winnipeg, Manitoba.

Strang, G. W. and Fix, G. (1973). "An Analysis of the Finite Element Method." Prentice Hall, Englewood Cliffs.

Thomas, G. B., Jr. (1972). "Calculus and Analytic Geometry" (14th ed.). Addison Wesley, Reading, Massachusetts.

Unwin, D. J. (1975). "An Introduction to Trend Surface Analysis." Geo Abstracts Ltd., Univ. of East Anglia.

Wahba, G. (1981). Bayesian confidence intervals for the cross-validated smoothing spline. *Tech. Rep. No. 645,* Dept. of Statistics, Univ. of Wisconsin, Milwaukee.

Whittaker, E. T. and Robinson, G. (1944). "The Calculus of Observations" (4th ed.). Blackie and Son, London and Glasgow.

Zienciewicz, D. C. (1971). "The Finite Element Method in Engineering Science." McGraw-Hill, London.

Index